000414242

000414242

Plea
belo
Boo
an
Re
54

WITHDRAWN FROM STOCK
The University of Liverpool

DUE RETURN

1996

CANCELLED

For conditions of b

0414242 -✗

Plant Surfaces

JUNIPER, BARRIE EDWARD
PLANT SURFACES
000414242

QK725.J91

Plant Surfaces

B. E. Juniper
Botany School, University of Oxford

C. E. Jeffree
Botany Department, University of Edinburgh

Edward Arnold

© B. E. Juniper and C. E. Jeffree, 1983

First published 1983
by Edward Arnold (Publishers) Limited
41 Bedford Square, London WC1 3DQ

All rights reserved. No part of this publication may be
reproduced, stored in a retrieval system or transmitted
in any form or by any means, electronic, mechanical,
photocopying, recording or otherwise, without the prior
permission of Edward Arnold (Publishers) Limited.

British Library Cataloguing in Publication Data

Juniper, B. E.
    Plant surfaces.
    1. Plant cells and tissues
    I. Title   II. Jeffree, C. E.
    574.8'2   QK725
    ISBN 0-7131-2856-9

Printed in Great Britain by
Butler & Tanner Ltd, Frome and London

Contents

001188

Preface

The plant cuticle, that delicate skin over the cellular tissue, was first observed 200 years ago. It was, as might be expected, the first plant surface examined in detail. But with a few exceptions, notably the pioneering biochemistry of A. C. Chibnall in the 1930s, the chemistry of this skin was too complex and its detail too fine for much advance to be made before gas–liquid chromatography and electron microscopy.

We now know something about the biosynthesis of the various polymers which make up the cuticle. Model systems can simulate its development. We know something too about the ways in which the environment can modify both this and other plant surfaces and the extent to which such surfaces are under genetic control. We know very little, as yet, of the ways the various chemical components of the plant's surface move from their synthesis within the living cells out to their point of deposition in or on the plant skin.

A cuticle is found over every young aerial plant surface, coating even the apical tip beneath repeated layers of leaf primordia. It penetrates down into the sub-stomatal cavity, under the superficial leaf salt glands in salt-secreting plants and covers the enzyme-secreting glands of insectivorous plants. It reaches its most massive development in the thick, commercially-valuable wax layers on the leaves of the carnauba palm, *Copernicia cerifera*. The epicuticular wax of such a plant surface was man's first plastic. It reaches its peak of subtlety in the delicate detachable wax scales of the trap mechanism of the *Nepenthes* pitcher from which no arthropod can climb. In its routine role it protects the plant from desiccation, hinders the penetration of pathogen and insect and may reflect, diffuse or concentrate the rays of the sun.

Under intensive agriculture, more and more substances, damaging and beneficial, are thrown or fall on to leaf surfaces. Amongst the former are weed-killers and the fall-out from industrial and nuclear activity; amongst the latter are insecticides, foliar fertilizers and irrigation water. Recently it has become apparent that almost every plant surface not only absorbs materials, but also discharges them. In fact the whole plant's surface, from apical tip to root cap, is an oozing, flaking layer. Sometimes, as in the root cap, this appears to be a highly organized and purposeful exercise. In others, such as most leaves, it is not at all clear how the plant benefits from this loss, but an often vigorous epiphyllous flora and fauna obviously does. Where the animal carries its alien flora, generally of a beneficial kind, on the moist warm walls of its gut, the plant does likewise, but on the exposed surfaces of root and shoot.

Many leaves, through leaching, are rich sources of nitrogenous compounds, polyphenols and salt solutions. They may discharge, into the atmosphere, heavy metals concentrated from the soil and also a range of hydrocarbons. Much of the coal and oil beneath the earth's surface may derive from this source. Young root surfaces may discharge into the soil a range of polysaccharides and some proteins, some of which may be signalling compounds to parasites or symbionts.

Any plant surface may become the home of a wide range of parasites or symbiotic organisms. Some epiphyllous plant species may have no other home than the surface of a tropical leaf and nitrogen-fixing micro-organisms on leaf or root are widespread.

The plant surface too is the recognition layer for friend or foe. It is the benevolent and compatible surface of the pollen grain or graft interface or the hostile surface of fungal hypha or *Cuscuta* haustoria.

From the inert, enveloping skin of the eighteenth century the plant surface is now seen as a dynamic adaptable envelope, flexible in both the import and export of materials, forming both an ecosystem in its own right and the first barrier between the moist, concentrated, balanced cell and a hostile ever-changing environment.

Oxford and Edinburgh, 1983   B.E.J. and C.E.J.

1 The Historical Perspective

'The structure of leaves is very simple. It consists of an outer skin or cuticle which is full of pores, the upper surface being varnished as it were. Cellular tissue is seen when the cuticle is removed' from *The Library of Agricultural and Horticultural Knowledge,* BAXTER, J.(1834). Lewes.

Over 300 years ago Robert Hooke, who first used the word 'cell' in a description of plants, was also the first to describe in detail the leaf surface and the hairs on a nettle (*Urtica dioica*).The presence of a superficial non-cellular membrane was first suggested by Ludwig in 1757 and Brongniart (1830–34) isolated this membrane which he named the 'cuticle' (Fig. 7-5), by allowing a cabbage (*Brassica*) to rot in water. Henslow (1831) detached a similar skin with nitric acid from the corolla, stamens and style of the foxglove (*Digitalis purpurea*) and Brodie in 1842 detected cuticles in fossil plants. The mid nineteenth century was a period of vigorous activity into the physiology of leaves and 'cuticular' versus 'stomatal' transport of gases was fiercely debated. Darwin was aware of the permeability of the cuticle over the glands of insectivorous plants (1875) and it was apparent as early as 1845 that this membrane was not only external but lay within, for example, the sub-stomatal chamber (von Mohl).

Ideas that are still the subject of debate have long histories. Could the plant surface wax pass out through the cuticle in a solvent carrier? This was suggested by Wiesner as early as 1871. De Bary (1884), in a monumental work on the classification of plant surface waxes, proposed that they migrated through cuticular canals; the suggestion is often repeated in modern literature. Apart from De Bary's description of almost every known waxy plant surface (transmission and scanning electron microscopy have only added detail to his major work), he also investigated commercial waxes, including carnauba (see section 7.7) and experimented both on their chemistry and biosynthesis. Lee and Priestley (1924) brought the subject well into the twentieth century with widespread investigations into the role of the environment in the development of the cuticle and its components, its biosynthesis, physiological functions and ontogeny. Chibnall and Piper's pioneering and very extensive work on the chemistry of plant waxes in the late 1920s and 1930s, laid not only a groundwork for the major typing and distribution of plant waxes but also established many of the basic ideas of their biosynthesis. The ground was now prepared for the refined techniques of the post-war years, chromatography in all its forms, autoradiography and electron microscopy, to exploit and to illuminate.

2 Techniques for the Study of Plant Surfaces

A vast spectrum of techniques, ranging from the most simple and observational to those involving sophisticated apparatus, has been applied to the study of the plant surface. Most of these are routine and have broad application elsewhere and have likewise been described in detail elsewhere. Only a few, e.g. scanning electron microscopy and certain isolation and analytical methods for chemical components are given a relatively full treatment here, the others are dealt with briefly and reference is made in the reading list to a more extensive coverage.

2.1 Microscopical Methods

2.1.1 Polarized light microscopy

Plane-polarized incident light can show how large molecules are orientated within a structure. If the molecules are non-random then plane-polarized light will distinguish between the axes of the substance; its refractive index will differ according to which axis is measured. The early anatomists distinguished in this way between the isotropic, randomly orientated layers, e.g. the pectin and wax layers, and the anisotropic orientated layers, e.g. the cutinized layer and cellulose-rich wall layers of a plant's epidermis.

2.1.2 Staining for the light microscope

Conventional staining for the light microscope has added little to our knowledge of the structure of plant surfaces principally because of the lack of a precise reaction between, for example, the stains for pectin (ruthenium red is specific only for carboxyl groups) or phloroglucinol in HCl which does not clearly distinguish between suberin and cutin. The waxes of plants are chemically so diverse that no useful staining reaction exists for them *in situ*. But separated waxes may be selectively stained (see section 2.2.2).

2.1.3 Scanning electron microscopy (SEM)

SEM is the simplest and most direct method for high resolution microscopy of plant surfaces. The most usual operating mode, the emissive mode, provides chiefly topographical information, producing an image with an uncannily natural appearance. The effect is similar to a reflected light micrograph or macro-photograph, but with about 300 times the depth of focus obtainable from light images.

A modern SEM has an optimum resolution of about 10 nm with biological specimens compared to about 2 nm for transmission electron microscopy (TEM) and 250 nm for optical microscopy (Grimstone, 1976).

The SEM may be used to examine the surface of any small ($<$1 cm^3), dry, robust specimen. Non-conducting materials tend to charge up in the electron beam, causing beam deflections and image distortion. This can be overcome by coating the specimen with a conducting film of carbon and/or gold or gold/palladium alloy, techniques borrowed directly from TEM, which conduct charge to earth via the specimen stage.

Delicate plant specimens, such as primary roots which collapse under vacuum, are usually dehydrated by critical point drying (CPD). The specimens are first fixed and dehydrated in a water/alcohol series and the alcohol replaced with acetone, amyl acetate or Freon 113. The solvent is then in turn replaced with liquid carbon dioxide in a pressure bomb. Finally the pressure bomb is sealed and its temperature raised to the 'critical point' (31°C), at which carbon dioxide undergoes a change of state, from liquid to gas, at the *same* volume. The gaseous carbon dioxide may be bled off and the specimen is thereby dried without having passed through a gas/liquid interface.

CPD gives excellent results with many tissues, but the solvents used may destroy plant epicuticular waxes. Better results are often obtained if the surfaces of plants are mounted fresh as untreated tissues and examined immediately using a low voltage beam to minimize charging, or briefly coated with gold in a sputtering device. Rapid handling and observation are essential and specimens cannot be stored. Delicate tissues, such as rose petals, which must be examined in an untreated state can be frozen in liquid nitrogen and viewed, still frozen, on a cryogenic specimen stage. This method, though technically demanding, provides a very high standard of tissue preservation free from artefacts of dehydration and solvents (Fig. 4-3).

Dry plant materials, seeds, pollen or wood which present no special problems of dehydration may be examined coated or uncoated as desired.

The SEM beam causes X-rays to be emitted from the specimen which are characteristic of the element which emits them, and an image can be constructed by an X-ray spectrometer of the distribution of an element in the scanned area of the specimen. X-ray emission mode microscopy has been used to investigate the chalk-secreting glands of *Plumbago capensis* and the deposition of silica in the tissues of *Equisetum*, rice (*Oryza sativa*) and other cereals. The cathodoluminescent mode can be used to visualize the distribution of fluorochromes which emit light in the electron beam. Various modes can be used in combination to give different types of information about the same area of the specimen surface. Because the SEM image is constructed from a linear sequence of information it is ideally suited to digitization and image processing. At the simplest level this permits electronic control of the image quality – contrast, and edge definition. More sophisticated analysis enables objects to be identified, counted and measured and it is possible to record the image, using video recorders, for subsequent computer analysis.

2.1.4 *Transmission electron microscopy (TEM)*

TEM made its first major impact on the study of the plant surface through the indirect replica method (Fig. 2-1) (Juniper and Bradley, 1958). This technique permits almost any surface, however delicate or opaque to be examined. Carbon, the commonest replica material, has virtues of coherence, low electron contrast,

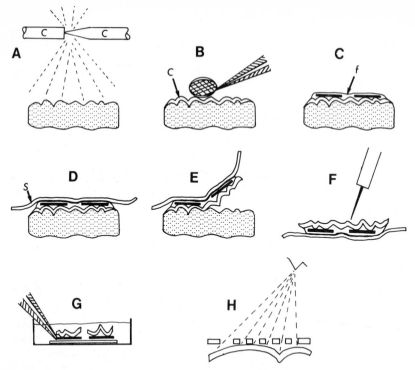

Fig.2-1   The carbon-replica technique. **A** The fresh plant surface is coated, in a vacuum, with evaporated carbon from an arc (C). **B** Grids dipped in Formvar are placed on the carbon-coated surface, which is then (stage **C**) flooded with 2% Formvar in chloroform (F). **D** As soon as this is dry a layer of sellotape (S) is placed over the Formvar layer, mechanically stripped (stage **E**) and the perimeter of each grid (stage **F**) is scored around with a needle. **G** This four-layered sandwich is placed in a bath of chloroform to remove the Formvar, the carbon replica is lifted away on the grid, inverted and obliquely shadowed, through the grid bars (stage **H**), with an evaporated heavy metal such as platinum.

conductance, the ability to replicate re-entrant angles and to replicate biologically moist surfaces. A good carbon replica (Figs 2-2, 2-3 and 3-2) can resolve 2 nm and combined, as here, with metal-shadowing reveals a wealth of structure well below the resolution of the light microscope. However, the techniques are tedious and suffer from the disadvantage present in all TEM work that the third dimension is suppressed or almost eliminated. It has, therefore, for almost all problems been supplanted by the SEM (Figs. 2-2, 3-7, 4-3, 4-7 and 5-2) (Baker and Holloway, 1971, Parsons *et al.*, 1974).

Difficulties of adjusting embedding materials to cope with the preservation of structures as diverse as cellulose, cutin and wax delayed the application of ultra-thin sectioning to plant surface studies. Many of these difficulties have now been overcome and the whole battery of minor techniques associated with TEM, including negative staining, freeze-etching, radioactive and inert X-ray analysis, ultra-histochemistry and high-voltage TEM, are now used.

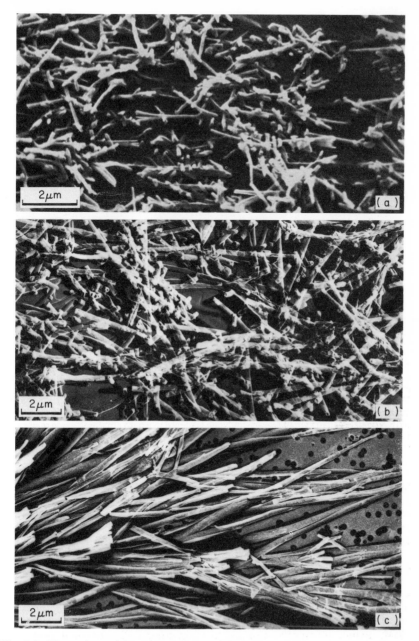

Fig. 2-2 Three views of the same surface. **A** SEM of Sitka spruce (*Picea sitchensis*) leaf surface showing epicuticular wax tubes. **B** As above, but TEM platinum-shadowed carbon replica. **C** Sitka spruce wax crystals washed from the leaf surface and recrystallized *in vitro* from solution in hexane. Gold /palladium shadowed carbon replica. TEM.

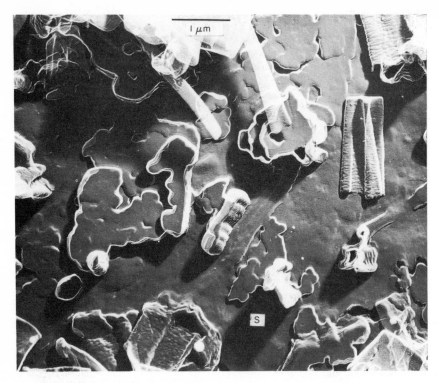

Fig. 2-3   A platinum-shadowed carbon replica of a cabbage (*Brassica oleracea*) leaf surface. (S) is the 'shadow' formed by the exclusion of platinum from the carbon surface.

2.2   Chemical Methods

2.2.1   *Isolation of the epicuticular waxes*

Plant surface waxes are chiefly composed of long chain aliphatic hydrocarbons with chain lengths between C20 and C35. The molecules vary in polarity depending on the position and kind of any substituted groups present, e.g. carboxyl ($-COOH$), hydroxyl ($-OH$) (see Figs 2-4 and 7-3). Most of these compounds are readily soluble in organic solvents. The non-polar n-alkanes dissolve easily in non-polar solvents such as hexane or benzene, but are much less soluble in polar solvents such as acetates and alcohols. The converse is true with the very polar carboxylic acids.

Since plant waxes contain a mixture of polar and non-polar constituents, solvents must be carefully chosen if they are to dissolve a representative sample of the wax constituents from the plant surface. Chloroform appears to introduce the least selective bias and it is the most generally used solvent for isolation of plant epicuticular waxes.

Most of the epicuticular wax dissolves from plant surfaces in the first 10 to 30 seconds of immersion in chloroform at 20°C. Successive brief washings in fresh solvent remove cuticular wax constituents from progressively deeper locations in the cuticle. These differ qualitatively from the epicuticular wax, but there are many compounds in common. Continued solvent extraction removes more of the lipid constituents of cells and fewer cuticular lipids, and for this reason intact, undamaged leaves are used. The wax can be extracted from separate surfaces of a broad leaf by washing them separately in a stream of chloroform from a burette. More local wax extraction can be achieved by swabbing plant surfaces with cotton swabs soaked in chloroform.

When sufficient wax is present it can be removed physically by scraping, or wiping with dry cotton swabs. In exceptional cases the wax layer may be thick enough to remove with tweezers, as in leaves of the carnauba palm (*Copernicia cerifera*). Useful information about the distribution of wax constituents in the epicuticular and cuticular layers comes from the comparison of wax obtained by physical means with samples obtained by different methods of solvent extraction.

2.2.2 Thin layer chromatography (TLC)

Early analytical methods needed long and tedious chemical procedures and large amounts of material for separating waxes into their constituents. Modern methods rely on chromatography for separation of the constituents and on sensitive methods which permit unequivocal identification of trace amounts of individual compounds (Holloway and Challen, 1966).

Thin layer chromatography (TCL) is a simple and sensitive method of separating wax constituents by constituent class. In this method a thin layer (0.25 mm) of absorbent, usually silica-gel, is spread on a glass plate. Spots of wax solution are applied to the surface of the plate with a glass capillary, and a solvent front is allowed to ascend the plate from one edge, carrying wax compounds with it. The distance travelled by a compound depends on its polarity and the polarity of the solvent. Chloroform solubilizes wax constituents so uniformly that most are carried to the solvent front. The least polar constituents are best separated with pure benzene, (Fig. 2-4), and the more polar constituents are resolved by a 60/40 mixture of chloroform and ethyl acetate. Details of staining reactions and applications of TLC for natural waxes are described by Holloway and Challen (1966).

2.2.3 Gas–liquid chromatography (GLC) and mass spectrometry (MS)

GLC is a powerful analytical method, enabling the general classes of wax compounds to be further resolved into their individual homologues and enabling quantitative estimations to be made. In GLC the stationary phase could be a hydrocarbon oil or silicone film on the surface of a powdered inert solid such as firebrick or diatomaceous earth, packed into a stainless steel column. The mobile phase is an inert gas, usually nitrogen or helium. The minimum quantity of a compound detectable using this system is about 0.01 μg.

Comparison of the behaviour of unknown compounds with known standards gives much circumstantial evidence of the identity of compounds, particularly when coupled with information about staining reactions from TLC. Such evidence can be further reinforced by derivatization of a compound which often results in characteristic shifts in retention and by information from infra-red absorption spectroscopy. However, only mass spectrometry (MS) is capable of unequivocal identification of compounds and it is used as the final arbiter. As a compound is eluted from the GLC column it passes into the MS where it is ionized by an electron beam. The molecules fragment into characteristic

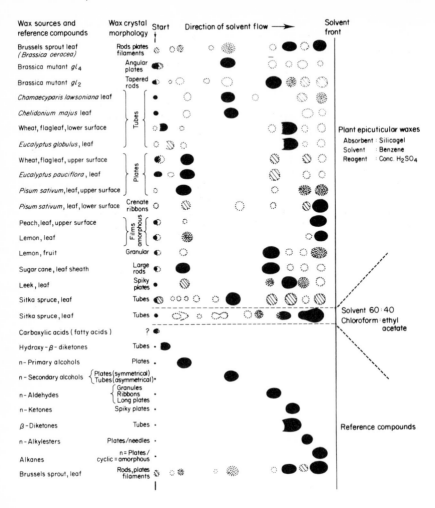

Fig. 2-4 Thin layer chromatogram of plant epicuticular waxes compared with known standards and correlated with wax crystal morphology.

patterns of ions, often including the parent ion, the positively-charged parent molecule. This ion 'fingerprint' can identify and quantify mixtures of isomers which are inseparable using GLC alone.

2.2.4   Isolation of plant cuticular membranes

The cuticle is strongly resistant to chemical attack and may survive natural decay long after the cells have gone, particularly in acid anaerobic environments. Early workers found that they could isolate plant cuticles by treatments as diverse as digestion in dilute nitric acid and allowing leaves to rot in water (Fig. 7-5). More recently cuticles have been isolated with mixtures of nitric and chromic acids, concentrated sulphuric acid and saturated solutions of zinc chloride in hydrochloric acid (Holloway and Baker, 1968).

Gentler methods are used for critical analytical work and it is now common practice to use enzymes. Mixtures of cellulase and pectinase (2–4%, pH 4.0, 37°C) remove the cell walls and detach the cuticle by dissolving the pectin (Fig. 3-4) which bonds it to the epidermis. Similar results are obtained with enzymes from *Helix pomatia* gut, or by incubation in the rumen of fistulated animals.

Mixtures of 2% ammonium oxalate in 0.5% oxalic acid, solutions of hydrogen peroxide, and EDTA have also been used and yield cuticles of the same weights as those obtained with enzymes.

The cuticles of many species resist isolation using these gentler methods and *Fragaria* cuticles fragment badly making them difficult to handle and measure. Holloway and Baker (1968) found that in difficult cases they could still isolate cuticular membranes with zinc chloride/hydrochloric acid mixtures. Conifer cuticles resist any of these methods, although they are isolated naturally by fungi and arthropods in leaf litter, and can be found there in considerable quantities.

3 The Anatomy, Ultrastructure and Biosynthesis of the Plant Surface

3.1 Lower Plant Surfaces

Unicellular algae, unlike the protozoa, have a wall of cellulose and pectin. As these algae evolved they aggregated into colonies embedded in mucilage, a chemical extension of the wall.

A seaweed is a relatively massive accumulation of cells with a thallus many cells thick, but it still has this mucilaginous coating. The only permanent slimy layer in a higher plant is that of the root cap which, incidentally, in maize has fucose as one of its mucilage sugars. This slime layer holds large quantities of water, resists loss by evaporation and assists the algal invasion of the littoral zone between the tide marks, where the plants are exposed twice daily to dehydration. Its lubrication also reduces the damage by the rubbing of one plant on another.

The seaweeds, except those briefly above high tide, have no drought problem, and do not hinder the movement of water through the plant surface. But land plants, particularly those in the drier habitats, have to conserve water all the time. In the higher plants the gas exchange surface is not, as in the algae the outside, but the mesophyll cell surfaces, loosely-packed photosynthetic cells with air spaces between and gas diffusion around them. The mesophyll is covered by the epidermal cells which are in turn coated with the water-impermeable fatty cuticle, separating the mesophyll from environmental contact. The light-transparent epidermal cells are interrupted by the stomata. These act as ventilators and allow the mesophyll cells to photosynthesize in a bright environment under controlled conditions of humidity and CO_2 concentration.

The evolution of this arrangement is obscure, but in the liverworts such as the Marchantiales, there are large, permanently open pores leading to internal photosynthetic tissues (Fig. 3-1), which may represent an intermediate state. A similar arrangement is known in fossils of primitive vascular plants such as *Spongiophyton*. The cuticle too plays an important part in the evolution and differentiation of land plants. It is the final barrier to water loss and must have been an early development in evolution as the thalloid layer of the early plant form began to develop and thrust up into the desiccating wind.

A cuticle is always present in land plants. However, the vestigial, but occasionally waxy cuticle in most Bryophytes – moss leaves are usually only one cell thick – restricts them to wet habitats, or has brought about a tolerance to desiccation without permanent damage. A fern, intermediate between higher plants and Bryophytes in its degree of terrestrial adaptation, has a prothallus (gametophyte generation) lacking either epidermis or cuticle, thus restricting it to a moist habitat. The sporophyte generation however is vascular, has a

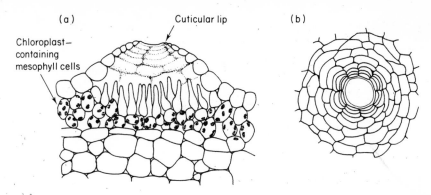

Fig. 3-1 Air pore structure in the liverwort *Conocephalum conicum* (Marchantiales). **A** Transverse section. **B** Surface view. A very thin cuticle covers the epidermal cells and forms a small lip around the pore which impedes the entry of water into the air chamber.

differentiated epidermis with stomata and a cuticle and is often much more tolerant of dry conditions.

3.2 The Epidermis in Higher Plants

Epi and *derma* are Greek for 'upon' and 'skin'. No epidermis forms over a root cap nor differentiates over a shoot meristem, but it is otherwise ubiquitous.

The *epidermis* arises from the outermost layer of the apical meristem in both shoot and root. In a shoot with a tunica and a corpus the epidermis forms from the outermost tunica layer. In a root the epidermis forms from the outermost cortical layer. Many plants, most annuals for example, retain the epidermis as long as they live. In others the epidermis of stems and roots is displaced with secondary growth by a periderm and sometimes, at a later stage, by a bark or a peel (Cutter, 1978). However, the root cap (section 7.1) may, in maize for example, replace every cell each day. The peripheral root cap cells do not form an epidermis, but are mucilage-secreting like the seaweeds.

Most epidermal cells are similar within a single species, but vary from species to species. Usually they are tabular in shape with various degrees of specialization related to their superficial position (see later). An epidermal cell fits tightly to its neighbours with no intercellular spaces (Figs 3-2 and 3-5). It may be elongated, as in many monocotyledons and some conifers (Fig. 3-5) or, as in *Pisum sativum* (Fig. 3-2), have a markedly wavy outline. Normally the only breaks are the stomata (Figs 3-3 and 4-1) and, more rarely, the hydathodes, structures adapted for water secretion. A normal epidermal cell's fine structure is undistinguished. Plastids are small, few in number, virtually colourless, (an interesting exception is *Phaseolus*) and unlike the guard cells (Fig. 4-1) these rarely contain starch. The epidermal cells, however, of a few water plants and ferns may contain fully functional chloroplasts.

The outer tangential wall of the normal epidermal cell is usually the thickest and may, as in many xeromorphic plants, be lignified. In some species these walls may accumulate calcium carbonate (the cystoliths) and in others, as in the *Gramineae, Cyperaceae, Palmae* and *Equisetaceae* (Fig. 3-3) grains of silica (plant opals). These opals grind down herbivores' teeth, can help the

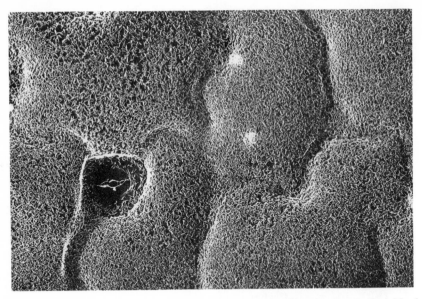

Fig. 3-2   Shadowed carbon replica (TEM) of a young pea (*Pisum sativum*) adaxial leaf surface. Wax crystals are ubiquitous except over the guard cells.

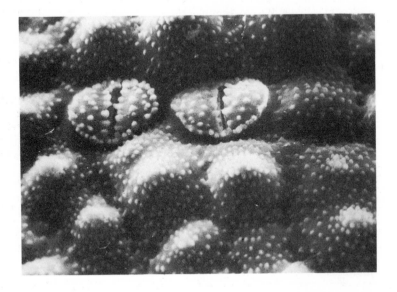

Fig. 3-3   Knobs of silica on the surface and stomata of *Equisetum*. SEM.

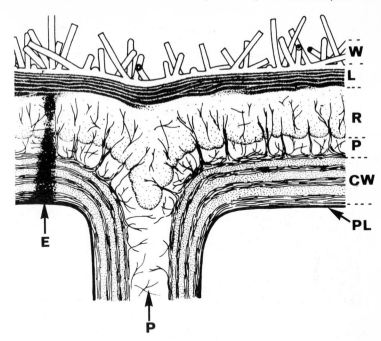

Fig. 3-4 The structure of the plant cuticle. E, ectodesma; P, pectin layer and middle lamella; PL, plasmalemma; CW, cell wall consisting of alternate layers of cellulose fibres and layers containing, predominantly, hemicellulose and pectin; R, reticulate region of the cuticle, cutin and wax tranversed by cellulose fibrils; L, lamellate region of the cuticle, separate lamellae of cutin and wax; W, the epicuticular wax.

archaeologist to identify long rotted plant material and, in *Equisetum*, are responsible both for its rough feel and former use as a pan scourer. Frequently beyond this wall, between the wall and the cuticle and joined to the middle lamella of the epidermal cells, is a thick layer of pectin or pectin-like material (Fig. 3-4). This layer is usually absent in rigid leaves like those of succulent plants. It may serve as a cushion between the segmented epidermis and the cuticular skin above. As the leaf flexes this biological sponge may help to absorb the differential strains. Beyond the pectin layer lie the layer or layers of the cuticle and, where present, the epicuticular wax (Figs 2-2, 2-3 and 3-2).

Structures known as *ectodesmata* are commonly found in the outer tangential walls (Fig. 3-4). Initially and erroneously called plasmodesmata they have no bounding membrane, do not join the cytoplasm of the epidermal cell below nor penetrate through the cuticle above. They cannot normally be seen in unstained preparations or by freeze-etching for the electron microscope. They are probably local reducing areas in the wall whose function is unknown, but whose frequency varies from day to night, place to place on a leaf and from almost any physical stimulus or environmental change.

Stomata are pores formed between two adjacent modified epidermal cells called the 'guard cells' (Figs 3-2 and 4-1). Subsidiary cells, modified epidermal cells adjacent to the guard cells, are often present as well (Figs 3-5, 3-6 and 4-1), and the cell group may be referred to as the stomatal complex. Stomata may be

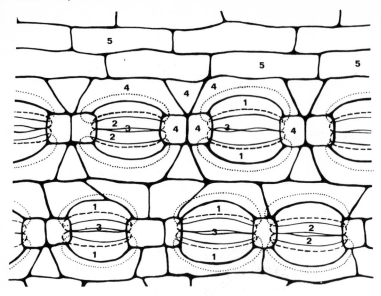

Fig.   3-5 The   pattern   of   epidermal   cells   in   *Picea   sitchensis.*
1 Subsidiary cells. 2 Guard cells. 3 Stomatal aperture. 4 Rhomboid, square and
triangular epidermal cells which form the packing between the stomata. 5 Linear files of
rectangular epidermal cells.

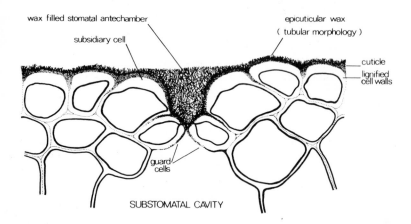

Fig. 3-6   The stomatal aperture of a leaf of *Picea sitchensis.*

raised above (Figs 3-3 and 4-2), at the same level or sunk below (Fig. 3-6) the
level of normal epidermal cells. In *Nerium, Hakea*, some *Eucalyptus* species and
some conifers (Fig. 3-6) the sub-epidermal stomata are sunk in pits or crypts
protected by hairs or waxy outgrowths (see section 4.1.2).

    Trichomes are structures formed solely from epidermal cells and adapted to a
variety of functions. Structures deriving from both epidermal cells, and sub-

epidermal layers are called emergences. The many types of plant hair, (Figs 3-7 and 4-4), scales, glands and some prickles arise from either source.

Trichomes may be single-celled, as in most hairs, or complex, as in secretory or absorptive glands (see Chapter 7). They may persist or be ephemeral. Usually they are of cellulose, but may be cutinized, occasionally lignified and more rarely impregnated with silica or calcium carbonate. Extreme examples are the cellulose-rich cotton hairs of commerce up to 7 cm long, the trigger hairs of *Dionaea* (section 5.4, Fig. 5-4) and the stalked glands of the carnivorous plants *Drosera* and *Pinguicula* (Fig. 3-8) which are up to several mm long, discharge mucilage, water and hydrolytic enzymes and absorb insect proteins. The looped hydrophobic hairs of *Salvinia* (Fig. 3-7) repel water from the leaf surface. Other glands secrete sugar (nectaries), terpenes, ethereal oils, balsam, resins, camphor and salts (Chapter 7), or hold pigments (Fig. 4-4).

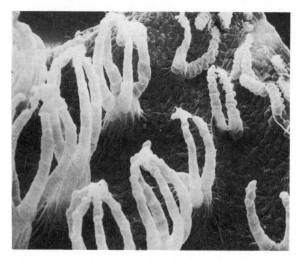

Fig 3-7   The hydrophobic hairs on *Salvinia auriculata* (SEM). Courtesy of Mr G. Wakley, Botany School, Oxford University.

A *periderm*, like a multiple epidermis, forms by periclinal divisions, usually in the epidermis or hypodermis. The zone of these periclinal divisions may persist as a cylindrical meristem, the phellogen, contributing cork cells both inside and outside. Cork cells are dead at maturity, have suberized cell walls and no air spaces. The cork layer may form as separate arcs giving rise to scaly bark or may form a complete cylinder, ring bark. Lenticels are often found; small areas in the cork made up of non-suberized cells with prominent air spaces. These pit-like areas in otherwise almost impervious cork, like some stomata, may have waxy outgrowths from the cells forming the side of the pit, *c.f.* Fig. 3-6, and these are thought to control water loss.

3.3   The Cuticle

All the aerial parts of a higher plant possess a cuticle. Cuticles are non-cellular and as such are distinguished from peels, rinds, or corks which are one or

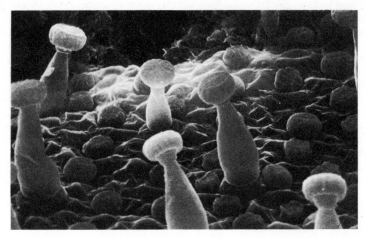

Fig. 3-8   The stalked gland on *Pinguicula vulgaris*. SEM. Courtesy of Mr G. Wakley.

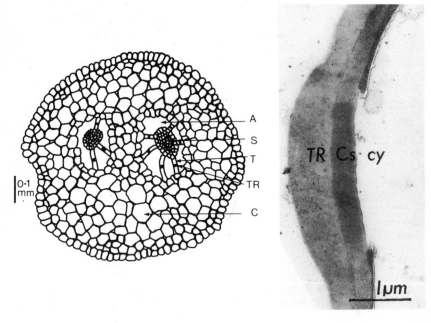

Fig 3-9   TS of a *Selaginella kraussiana* stem and TEM of the cuticle developing to different thicknesses on a trabecula cell. C, cortex; S, stele; A, air space; T, trabecula (a spoke-like cell); TR, trabecular ring, a thickened cuticular region adjacent to the Casparian strip (Cs); Cy, cytoplasm of the trabecular cell. Courtesy of Dr Barbara McLean, Department of Biochemistry, Imperial College, London.

more layers of more or less suberized cells. The cuticle is often multi-layered and its inner region merges into the cellulose of the epidermal cell wall (Fig. 3-4). It may even, as in the semi-porous glands of *Dionaea, Drosera, Drosophyllum* (section 5.4) be no more than the perforated incrustation of a thin cellulose wall.

Apart from bounding the epidermal cells of the leaf or stem the cuticle, in a thinner form, extends within and lines the sub-stomatal cavity (Fig. 4-1). It also covers the surfaces adjacent to the internal canals in certain Pteridophytes such as *Selaginella* (TR in Fig. 3-9). A cuticle is also found within other plant tissues, e.g. at the base of leaf salt glands (see Chapter 7 and Fig. 7-2), at the base of certain glands in insectivorous plants (Chapter 5), within some seed coats and within the water absorbing glands of some bromeliads such as *Vriesia psittacina*.

Commonly the upper and lower surfaces of the same leaf may have a different cuticle; there may also be epicuticular wax present on one surface but not on the other or local differences in waxiness on a single surface. Commonly plants grown under extreme conditions, e.g. high light intensity, low nutrient supply and wind exposure, will have thicker cuticles than the same species grown, for example, in a greenhouse.

As we shall see in Chapters 4 and 7 although the cuticle is almost universally present it is sometimes little or no barrier to water or other substances moving in or out.

The cuticle is generally glued tightly to the tangential wall of the epidermal cell below. There is, however, a gradation of tenacity which varies from plant to plant and from one part of a plant to another as we have seen in Chapter 2.

Plant wax is embedded in and sometimes exuded over the surface of the cuticle (Figs 2-2, 2-3 and 3-2). Not all plants have wax on their surface, the so-called epicuticular wax, but probably a majority do. Some plants have a prominent waxy bloom (Figs 2-3 and 3-2). This bloom is the reflection and scattering of light on the surface by wax crystals of diverse forms and angles, whose dimensions are close to or only slightly above the wavelengths of light. The bloom of the cabbage leaf (Fig. 2-3) is due to corrugated wax tubes about 1.2 μm long and about 0.5–1.0 μm in diameter. Other bloomed surfaces may be formed of crystals like needles, cornflakes, ribbons or macaroni. However, the surfaces of the bloomed fruits of the grape (*Vitis vinifera*) or plums and prunes (*Prunus domestica*) carry no more wax on their surface than many apples (*Malus sylvestris*), which when mature have no obvious bloom, simply a perceptibly greasy layer. Olive and lemon leaves, although very rich in wax, likewise have no obvious bloom. More wax can often be found on the lower (abaxial) leaf surface than on the upper. The hybrid tea rose leaf (*Rosa* sp.) is shiny and virtually wax-free above, but has a well-marked bloom below. A summary of plant wax chemistry is given in Chapter 7.

3.4   Sporopollenin

Sporopollenin, the main constituent of both spores and pollen exine (Fig. 6-1), spores and the walls of a few primitive algae, is one of the most resistant plant materials. It is found as early as the pre-Cambrian and pollen skeletons of immense age enable archaeologists to reconstruct changes in natural vegetation or to identify food crops on sites inhabited by early man.

Sporopollenin is a carotenoid polymer, a polyester of several monomers of which β-carotene and zeaxanthin are the most common. Its biosynthesis is not yet understood, but, at least in pollen exine, it is thought to be synthesized by collaboration between the haploid pollen surface and the maternal tapetal cells which surround it.

3.5   Origins and Biosynthesis of Pectin, Cutin and Wax

The *pectin* layer between the cellulose wall and the cuticle proper (Fig. 3-4) has not been isolated and characterized, but it is likely to be rich in α-1,4-linked D-galacturonic acid residues. The carboxyl groups on this polymer may be partially methylated into the cutin polyester. Enzymes are known which can carry out the interconversions UDP-D-glucose $\rightleftharpoons$ UDP-D-glucuronic acid $\rightleftharpoons$ UDP-D-galacturonic acid and hence back into the central carbohydrate pool of the cell. How the pectin gets into the cellulose/cutin interface is unknown, but analogies from other well-studied pectin exporting cells, such as those of the root cap, suggest strongly that the dictyosomes of the epidermal cells would be involved.

The origin of the *cutin* precursors in the epidermal cells is now known from recent studies of biochemical pathways. The unsaturated fatty acids, such as oleic and linoleic acids, are probably synthesized in the epidermal cells where they convert to hydroxy-acids before reaching the surface. Here they are changed to hydroperoxides by a lipoxidase enzyme. The hydroperoxide groups rapidly combine with themselves and other groups to form peroxide ($-O-O-$) linkages, a process which is accelerated by ultraviolet light.

A study of the origin of the *waxes* starts with Chibnall and his co-workers in the 1930s, who suggested that the C29 alkane, non acosane, could be formed from two C15 acids by head-to-head condensation, with the common C29 ketone, nonacosan-15-one as the intermediate compound. Although an attractive idea, the C15 acids needed for it to work were not known in waxes at that time. Kolattukudy (1970 and 1976) has now shown that the C16 acid, palmitic acid (Fig. 3-10) is incorporated intact into long chain alkanes without losing its

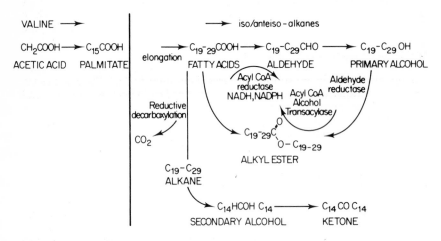

Fig. 3-10   The biosynthetic pathways of plant waxes.

carbonyl carbon atom as would be required in a head-to-head condensation. This suggests that long chain hydrocarbons form from shorter molecules. He has now found that labelled acetate units make even-carbon numbered long chain acids which are finally decarboxylated to odd-number alkanes. Palmitic acid's synthesis from acetate (C_2) is dependent on light and takes place in the mesophyll cells, probably in the chloroplasts, before being passed to the epidermis for final elongation and decarboxylation.

The intricate fine structure of the *epicuticular wax* (Figs. 2-2, 2-3 and 3-2) develops well away from the epidermal cells and it has not been clear until recently how the characteristically shaped structures develop.

De Bary in 1871 suggested that the wax is extruded through cuticular pores while in a semi-liquid state, rather like toothpaste from a tube. Wax microchannels were supposed to traverse the cuticle but there is no evidence that pores reach the outer surface of the cuticle at any stage of development. Moreover, no evidence exists for a massive pressure gradient across the cuticle, nor can extrusion explain a wax tube (Fig. 2-3) or wax crystals with a variable cross-section.

The precursors of cutin are secreted through the cell walls and are probably assembled under enzymatic control at the surface. Waxes are secreted through this membrane as it develops and it has been suggested that wax production is eventually prevented by the thickening of the cuticle, resulting in the blockage of pores through which the wax can migrate. However, the blockage of wax in this way should result in excess wax beneath the cuticle. Such accumulations of wax do not happen and probably the cuticle can be considered as a molecular mesh through which wax molecules, perhaps in solution in organic solvents or enclosed in protein shells, can percolate. The development of the wax layer, therefore, only stops when wax synthesis stops and can, as we shall see, start again after injury.

The morphology and composition of plant waxes can change during development. On apple fruits (*Malus*), where gradual expansion takes place over a long period, the ratio between esters and ursolic acid (a triterpenoid) increases during growth and in storage, particularly with varieties such as 'Bramleys', the surface of the fruit loses its crystalline bloom and becomes greasy from short chain esters. Early leaves of wheat and barley produce mainly primary alcohols while later leaves synthesize wax rich in β-diketones. Normally, however, the composition and morphology of the wax remains constant and it is possible to make some correlation between them.

Many commercial waxes such as shoe polish, beeswax, wax crayons, etc., develop, albeit slowly, crystalline surfaces if left undisturbed. Plant waxes too can undergo transition crystallization if melted and cooled in the laboratory and the crystals so produced resemble the epicuticular originals. With similar success plant waxes have recently been crystallized from solutions of organic solvents (Fig. 2-2) (Jeffree *et al.*, 1975).

Plant waxes, therefore, are able to organize themselves into crystalline structures independent of the cells or the cuticle, and their structure is very closely related to their chemistry.

The waxes of some species, e.g. some *Eucalyptus* species and *Chrysanthemum segetum* can be repaired to some extent when damaged, but usually only while the leaves are still expanding, not after they are mature. The amount of the regenerated wax generally cannot restore the wax layer to the thickness of undamaged areas, i.e. damage does not induce *de novo* synthesis but the 'regeneration' simply represents the remaining wax still to be synthesized at the time the damage occurred. The cuticle may also repair if it is badly damaged. Lipoxidase activity increses markedly when the cuticle is damaged (Chapter 5), and the lipoxidase apparently induces increased fatty acid synthesis.

Plants grown in high humidities produce more wax and have thinner cuticles than when grown normally. High temperatures also seem to induce more waxiness in most plants, although certain *Eucalyptus* species become glaucous only if subjected to mild frosts. However, the most profound determinant of wax production is light. Pea plants grown in darkness produce little wax on their leaves, but develop wax normally when transferred to full light. Most plants appear to need light intensities of at least 20% full daylight for normal wax synthesis (Juniper, 1960). The final elongation of the fatty acids in the epidermal cells is only light dependent insofar as it is limited by the supply of palmitate. This mechanism has a considerable adaptive significance since high violet/ ultra-violet intensities can be damaging and, as will be seen later (Chapter 4), the wax layer may have a light-protective function.

3.6   The Genetic Control of the Plant Surface and its Development

The fully mature leaf surface, with its lamellated cuticle and, where present, its epicuticular wax is the product of the interaction between the environment and the genetic make-up of the plant. Apart from light many other factors such as wind speed and chemicals in the soil, e.g. trichloracetic acid both in the laboratory and the field, can distort the surface pattern in a variety of ways. We must assume that the nature, thickness and surface features of the cuticle are under genetic control, but only the epicuticular wax provides easy genetic markers (von Wettstein-Knowles and Netting, 1976). The presence or absence of wax, modification of acicular shape or a shift in the proportion of alkanes are fairly easily detected, sometimes even to the naked eye.

Marked abnormalities in waxiness are easily spotted in plants in the field simply by sprinkling them with water. A more accurate assessment is given by such tests as the contact angle of water on the leaves (see Chapter 4), and changes in the wax ultrastructure, which commonly accompany these mutations, can be detected by SEM or TEM. Work began on *Zea* surface genetics in the 1930s and this was extended by the work of Bianchi, and his co-workers in the sixties from whose work Table 1 is taken. They confirmed the earlier work showing the dominance of glaucousness over 'greenness' (i.e. an absence of wax), but it only applies to the first five leaves of maize; older leaves are green regardless of genotype (Table 1).

There is no detectable difference between homozygous *Gl Gl* and heterozygous *Gl gl* genotypes and this seems common in wax genetics. Glaucousness is controlled by at least five genes: Gl_1 (located in chromosome 7), Gl_2 (chromosome 2), Gl_3 (chromosome 4), Gl^H (chromo-some 9) and Cg (chromosome 3).

On a pea leaf (*Pisum sativum*) (Fig. 3-2), over 50% of the wax is the *normal* alkane hentriacontane, along with a mixture of secondary alcohols also with 31 carbon atoms and comprising mainly 16-hentriacontanol. But four mutants have been found all having a wax with much less hentriacontane and hentriacontanol; the total alkane content goes down to 10%, the ester, aldehyde and primary alcohol proportion rises. The drop in alkane content and the reduction in glaucousness in the mutants confirms that highly crystalline surfaces usually have a high proportion of alkanes in their waxes.

Small changes in the alkane patterns of leaf surface waxes can also be used to identify the parentage of particular species of varieties of plants. The genus *Cupressus* is not native to East Africa, but several varieties of *C. lusitanica*, originally from Mexico, are now of major economic importance. It was thought that these varieties might have arisen by hybridization with the Californian *C. macrocarpa*, also widely planted there. But by analysing the leaf surface alkanes the successful Kenyan strain of *C. lusitanica* is shown to have no chemical affinity with *C. macrocarpa* and is almost certainly a locally derived mutant (Dyson and Herbin, 1968).

Table 1 'Green' mutants of maize (*Zea mays*) leaves.

Gene	Gl-	gl_1gl_1	gl_2gl_2	gl_3gl_3	gl^Hgl^H	$cgcg$
Leaf 1	fully glaucous	green	slightly glaucous with small projections	slightly glaucous with small projections	fully glaucous	mixed glaucous/ green
2	,,	,,	,,	,,	,,	,,
3	,,	,,	,,	,,	slightly glaucous with small projections	,,
4	,,	,,	green	,,	,,	,,
5	slightly glaucous	,,	,,	green	,,	green

4 The Plant Surface in Action

4.1 Interface with the Environment

4.1.1 The general transpiration/gas exchange problem

Land plants have conflicting needs. They must offer a large surface area to the sun and atmosphere to intercept light and absorb carbon dioxide, but keep the exposed surface area small to reduce water loss. Enclosing the mesophyll in a waxy cuticle was an important step in conquering the aerial habitats. Gas exchange could thereby be restricted to specified openings (stomata) in the surface. Sensitive guard cells evolved later permitting the stomata to open or close in response to the environment and bringing gas exchange under control.

Plants are very largely water and lose water continually by evaporation. Ordinarily the loss is balanced by uptake of water from the soil, its rate is conditioned by the properties of the plant surface, i.e. its chemical composition, dimensions, fine structure and gross morphology. The plant surface is also the interface with a thermal environment with which it exchanges energy. Heat is gained by and lost from plant surfaces by radiation, convection and conduction. Water exerts a stabilizing influence on the temperature of plants by its capacity to absorb energy when it evaporates (latent heat of evaporation) and to give out energy when it condenses without change in temperature. The energy balance of plants through transpiration is, therefore, connected with the area exposed to the light, the surface area to volume ratio, the colour and reflectivity of the surface and its texture and shape.

4.1.2 Stomatal and cuticular transpiration

Non-vascular land plants, so far as we know, transpire more or less evenly over their whole surface. In higher plants with stomata and a cuticle there are two components to the transpiration pathway; a large but controlled stomatal component and a cuticular component, largely but not completely uncontrolled, some 10 to 40 times smaller. Their relative magnitude varies greatly between species, because of anatomical differences (Meidner and Mansfield, 1968).

The cuticle, despite its hydrophobic properties, is semi-permeable to water and has a finite water content, since water passes across it. However, both the pathway of cuticular water transport and the site or sites of the evaporation from the surface are unknown and it seems most probable that the permeability of the cuticle to water is molecular and not usually through large discrete pores or channels.

The guard cells around the stomata (Figs. 3-3, 3-6 and 4-1), alter the size of the

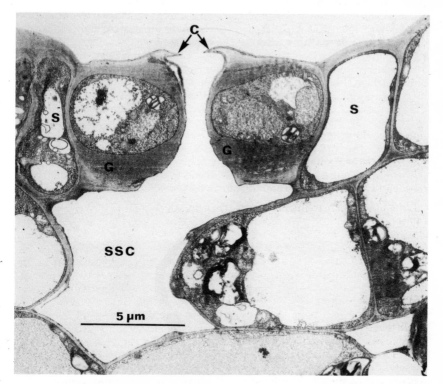

Fig. 4-1   Stoma, guard cells and sub-stomatal chamber in a lemon (*Citrus*) anther. (C) cuticular lips, (G) guard cells, (S) subsidiary cells, (SSC) sub-stomatal cavity. TEM. Courtesy of Dr E. Pacini, Department of Botany, University of Siena, Italy.

stomatal aperture in response to any environmental shift and influence water loss from the leaf. These shifts are chiefly changes in atmospheric humidity (saturation deficit) and leaf temperature, which is an expression of its energy balance. But stomata also react to light and CO_2 concentration in the mesophyll air spaces. Guard cells change shape through turgor changes and transient responses are often superimposed on a metabolically determined pattern. For example, the stomata of most mesophytic species have an endogenous, circadian cycle, opening at dawn and closing at night, even if the normal cycle of light and darkness is interrupted.

Since the function of stomata is to allow a high rate of CO_2 diffusion to the mesophyll and maintain optimum mesophyll turgor, it follows that the guard cells should lose water earlier than the mesophyll cells. Guard cells draw water indirectly, via the mesophyll and epidermal cell walls. They are, moreover, virtually isolated at maturity by extensive plasmodesmatal blocking. This isolation restricts the line of supply and thus the guard cells will close the stomata before the mesophyll cells become stressed. In conifers (Fig. 3-6) and in many xeromorphic species such as *Hakea* and *Nerium oleander*, the epidermis is multilayered and lignified, increasing this hydraulic resistance.

Stomatal transpiration is by molecular diffusion of water vapour. Its rate is inversely proportional to the size of a series of resistances to diffusion which occur between the moist mesophyll cell walls, where evaporation occurs, and the atmosphere. Resistances to diffusion vary with the dimensions and tortuosity of the system of internal air spaces, e.g. the spaces between the mesophyll cells, the substomatal air space (Fig. 4-1), the stomatal aperture, the stomatal antechamber and also the boundary layer of still, moist air which clings to the plant surface (Fig. 4-2). The volume of the internal air space, 5 to 40% of the internal volume of leaves, is reduced by 10 or 20% in wilted plants, increasing the resistance of that component of the water vapour diffusion pathway. At the same time, hydraulic resistance to water flow in the mesophyll cell walls increases, limiting the supply of water to the sites of evaporation, which retract deeper into the wall structure. In liverworts like *Conocephalum* (Fig. 3-1) such mechanisms are probably the chief means of transpiration control. The diffusive resistance of wilted plants increases about four-fold. In higher plants the mesophyll resistance for water vapour is very small, perhaps 0.1 to 1.0 s cm^{-1} and since the resistance of closed stomata is much greater, of the order of 40 to 120 s cm^{-1}, changes in mesophyll resistance are of little significance in controlling transpiration.

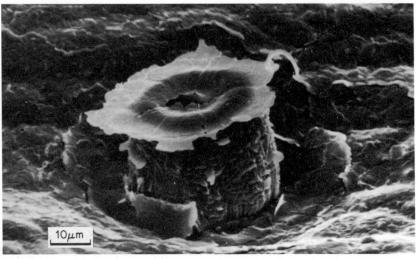

Fig. 4-2   Raised stoma on the xerophyte *Euphorbia tirucalli*. SEM. Courtesy of Dr W. Barthlott, Institut für Systematische Botanik und Pflanzengeographie, der Universität, Heidelberg.

Beneath the stomatal aperture there is often a pronounced substomatal cavity, of low resistance, a meeting-of-the-ways for the various mesophyll air spaces (Figs 4-1 and 4-3). The guard cells of plants from wet environments meet along a knife-edge contact face, as in the duckweed *Lemna*, and when open offer the smallest possible resistance for a pore of that diameter. In many mesophytes, and especially in plants adapted to dry environments, the xerophytes, the depth of the stomatal aperture is increased by cuticular ridges above or below the guard cells, (Figs 4-1, 4-2 and 4-3), or by overarching of the stomatal accessory cells to form an external stomatal antechamber. Such structures must increase the diffusive resistance of the stomata and may be xeromorphic adaptations for the control of water loss. However, interpreted in this way alone they are a paradox. The resistance of the closed stoma is very large, while transpiration from an epidermis with fully open stomata can, surprisingly, approach that from a free water surface of the same area, despite the fact that only about 2% of the surface is available for transpiration. The stomata are, therefore, efficient regulators of water loss and it appears counter-productive to hold stomata open while simultaneously reducing their conductance with obstructions.

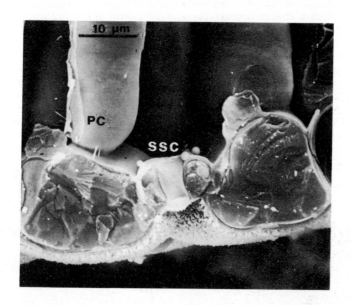

Fig. 4-3 An abaxial stoma on a *Lupinus albus* leaf prepared by the sputter-cryo method and photographed in the SEM. (SSC) sub-stomatal cavity, (PC) palisade cell. Courtesy of Dr A. Beckett, Department of Botany, Bristol University.

However, the ridges (Fig. 4-2) may provide a 'Venturi-effect' circulation like prairie-dog holes, when the stomata are open. The ridges may also ensure the optimum balance between transpiration control and carbon dioxide absorption. The resistance of the pathway for carbon dioxide from the free atmosphere to the chloroplasts is much greater than the pathway resistance for water vapour from the mesophyll surfaces to the atmosphere. This is because the mesophyll cells offer a large resistance to CO_2 diffusion in the liquid phase. Any small resistance around the stomata therefore reduces transpiration considerably, but reduces CO_2 absorption, which is already limited by the mesophyll resistance, by much less. For example, in Sitka spruce (Fig. 3-6) the wax-filled stomatal antechambers reduce transpiration by about two-thirds but reduce CO_2 absorption by only a third. In other examples of wax-filled stomata it has been suggested that this may be a device to prevent the flooding of the chambers where the leaves are commonly wetted by rain or dew, or to prevent the ingress of hyphae.

4.2   The Boundary Layer

Frictional drag slows down the air at plant surfaces, forming a boundary layer of still air up to several millimeters thick. Transport of gases across the boundary layer is by molecular diffusion and has diffusive resistance. In still air this can limit the maximum rate of transpiration out of fully open stomata. Air is rarely still and at air speeds above about 1–3 m s^{-1} the boundary layer is thin and allows stomatal control over the whole range of opening. The morphology of many plants influences the magnitude of the boundary layer and it is tempting to look for adaptive significance in specific features. At a given wind speed the boundary layer will be thickest on large leaves with entire margins, or on those with roughened, hairy or papillose surfaces (Fig. 4-4) which increase frictional drag. Narrow leaves, needles and broad leaves with toothed margins will encourage turbulence and thus thinner layers. However, the ecological significance of such features is difficult to determine. In strong sunlight leaf temperatures may rise several degrees above ambient and evaporative cooling can be important. In their natural habitat the very large leaves of banana, *Musa* sp., generally fray along well-defined lines of weakness, but can be damaged by overheating if kept intact, when the boundary layer probably limits transpiration and evaporative cooling. In some species with hairy leaves, the diffusive resistance of the boundary layer trapped by the hairs may have much the same physiological significance as that of stomatal antechambers and cuticular ridges and it is therefore interesting that the microphyllous leaves of

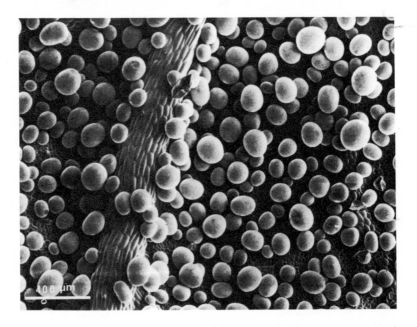

Fig. 4-4   The adaxial surface of a young leaf of *Chenopodium amaranticolor*. Balloon-shaped trichomes full of magenta sap cover the whole surface except the vein. These presumably have antitranspirant and boundary-layer-thickening effects, but the role of the pigment is unknown. SEM.

conifers, with thin boundary layers, have especially well-developed anti-transpirant antechambers (Fig. 3-6). In other species with hairy leaves the trichomes may have more to do with increasing the reflectivity (see section 4.4) of the leaf surface than with encouraging a thick boundary layer, emphasizing the pitfalls of ascribing a single function to a morphological feature.

4.3 Xerophytes

Although many lower plants, notably the algae, lichens and some Bryophytes, can survive repeated desiccation, only a few vascular plants can do so. Nevertheless, vascular plants survive in environments where water is so scarce that non-vascular plants cannot survive. This penetration of dry habitats is due largely to the evolution of a vascular system, enabling water to be scavenged from deep in the soil and to be redistributed among the aerial parts of the plant. But it is due also to the development both of a waterproof cuticle and stomatal control to restrict water loss.

Waxes on both plant and insect surfaces waterproof and protect. Cuticular wax waterproofs the plant so effectively that only a very small proportion (<10%) of the water lost by a plant passes through the cuticle (Schönherr, 1976 Sutcliffe, 1979). There is, however, no clear association between surface wax and xerophytic success and this has been interpreted as indicating that wax has little survival value. However, if wax is prevented from forming or removed from a plant surface transpiration rates increase several fold. Work with artificial membranes suggests that the aliphatic compounds, e.g. the hydrocarbons, alcohols, aldehydes and wax esters are far more effective at reducing transpiration than the alicyclic compounds, e.g. the triterpenes.

Plants which are specially well adapted to dry habitats are known as xerophytes, from the Greek *xeros* = dry and *phyton* = plant. Xerophytes adopt a wide range of strategies for survival in dry conditions and the distinction between the ordinary run of water conservation and xerophytism is a fine one. Plants which grow and flower in conditions too dry for mesomorphic species usually have specialized xeromorphic anatomical features. A thickened cuticle and wax layers are common xeromorphic features. Often the epidermis is thickened as well, or multilayered and the cell walls may be heavy and lignified or cutinized to reduce their permeability to water. In Scots pine the lignified cell walls of the hypodermis fill the cell lumen, and presumably impede water loss. Similarly thickened epidermal and hypodermal layers occur in *Nerium*, *Hakea* and marram grass (*Ammophila arenaria*).

Stomata of most land plants show some degree of xeromorphy *c.f. Lemna* sp. In land plants the contact faces are almost always thickened (Figs. 4-1 and 4-3) and the pathway through the stomatal aperture may be restricted by cuticular protruberances. In Scots pine and *Hakea* the stomata are overarched by extensions from the accessory cells, forming an antechamber which, in conifers, is usually wax-filled, further increasing its diffusive resistance. In *Nerium* (Fig. 4-5) the stomata are grouped together in hair-lined pits or crypts in the leaf surface.

The cuticle is never perfectly impermeable to water and the surface area exposed to the atmosphere relative to the internal volume of tissue is therefore reduced in most xerophytes. Some xeromorphic plants have a variable surface/volume configuration. Plants such as *Festuca glauca*, *Ammophila* and *Erica* (Fig. 4-6) can roll their leaves when stressed, enclosing the stomata in a tube

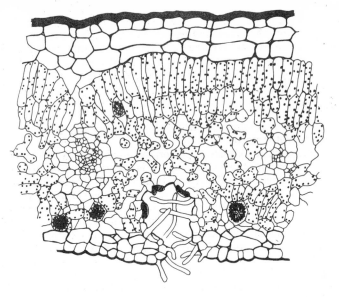

Fig. 4-5   Line drawing of a t.s. of a leaf of *Nerium oleander*. The stomata on the abaxial surface are deeply sunk in crypts and trichomes traverse the crypt and the crypt aperture.

lined with protective hairs and reducing the exposed surface by > 50%. Extreme reductions of surface area are found in the cacti and the ecologically equivalent *Euphorbiaceae* with thickened, succulent leaves or leaves reduce to spines. The photosynthetic areas are the fleshy stems. The ideal shape in which the surface area to volume ratio is a minimum is the sphere and some cacti, notably *Mammillaria*, approach this shape closely. The reduced leaf spines and glochids often form dense felted mats which reflect light. They help to reduce the thermal load and prevent overheating and at the same time both restrict air movement close to the plant surface and trap a boundary layer of moist air.

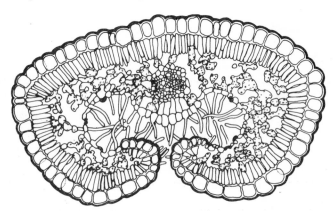

Fig. 4-6   Line drawing of a t.s. of an *Erica* leaf. Under dry conditions the leaf rolls and creates a longitudinal crypt similar, on a larger scale, to that seen in Fig. 4-5.

The South African Mesembryanthemaceae have thin cuticles despite inhabiting xerophytic niches. In these species greatly enlarged epidermal cells protrude from the leaf surface like balloons. In specimens growing under water stress, these 'balloon cells' are larger and more numerous, overhanging the epidermis and stomata in their vicinity. These cells appear to be able to absorb water from humid air. They can also alter their shape, flattening under stress so that the maximum area of epidermis is covered, and becoming more upright when turgid, so that gas exchange is less restricted. They thus behave rather like a second order of stomata, controlling the gas exchange pathway resistances in response to water status.

Some species with extreme xeromorphy inhabit moist environments, but occupy niches where the water supply is intermittent or precarious. The epiphytes of tropical rain forests include *Tillandsia* and other bromeliads, which, certainly as a group are almost equally at home in savannah. *Tillandsia* is specially interesting, because its roots are reduced or absent. Most bromeliads have water-absorbing cells on their surfaces (Fig. 4-11) and as we shall see in section 4.6 may absorb water from rain or dew directly into the leaf.

The effect of felted layers of hairs such as those on leaves of *Stachys lanata*, *Banksia stellata* and *Verbascum* and Fig. 4-4 on reducing transpiration was demonstrated by Haberlandt almost a century ago. As well as reducing thermal load by reflection they almost certainly reduce heat loss at night. The giant arctic alpine *Senecios* of African mountains cope with a diurnal temperature cycle of $-10°C$ to $+30°C$. They conserve heat at night by enclosing the apical rosette in layers of leaves covered with silver reflective hairs and thus prevent the growing point from freezing. In *Espeletia*, cited by Haberlandt (1914), unbranched hairs rising vertically from the surface undergo synchronous twists so as to form parallel layers of felting, alternating with loose zones. This arrangement presumably gives protection against the violent desiccating winds which prevail in the Paramos of Venezuela where the plant lives. Peltate hairs and scales may also have xeromorphic significance. In *Rhododendron* peltate hairs on the lower leaf surfaces of the xeromorphic types cover and protect the stomata which are grouped beneath them. In other species the stomata occupy more exposed areas of the epidermis and the peltate hairs are less frequent or absent.

4.4   The Optics of Leaf and Petal Surfaces

4.4.1   Light absorption

Many plant organs are phototropic, i.e. moving in a given direction in response to a light stimulus, or photonastic, changing their posture with no particular directional relationship to the changing light source or intensity. The petioles of leaves and the bases of petals frequently respond in this way, but do not seem to possess special light-receiving organs. The photo-stimuli received seem to be transmitted by unknown mechanisms to remote motor cell regions which respond. Thus the leaf blade of *Begonia discolor* assumes a fixed heliotropic position even though the petiole is totally shaded. However the primary leaves of *Phaseolus vulgaris* will orientate correctly, even when covered, provided that the petiole is exposed.

Many adaxial foliar epidermises have outer walls which are more or less papillose while the inner walls are approximately flat and parallel to the leaf surface. Each normal epidermal cell will therefore act as a plano-convex or condensing lens. These 'velvety' leaves are common features of the plants of tropical rain forests, e.g. *Begonia rex, Ficus barbata, Cissus discolor* and *Adoxa moschatellina.* So marked are the papillae of each cell that they can continue to act as lenses even if more or less flooded with water. Some of these papillose epidermal cells may not only possibly function as lenses, they are even constructed in the same way. *Petrea volubilis* has a disc of silica embedded in the outer tangential wall just under the papilla.

In a few species, e.g. *Dioscorea quinqueloba* and *Fittonia verschaffeltii* (Fig. 4-7) most of the epidermal cells are conventional in appearance and interspersed between them are larger protruding dome cells, the ocelli. The surface is smooth and the cell virtually transparent and up to 200 of these specialized cells are found per mm². Such surfaces can detect differences in light intensity approaching 1%, i.e. similar to that of the human eye, but very little is known of the mechanism, except that it seems highly likely that phytochrome, a chromoprotein found in all plants, is involved at an early stage.

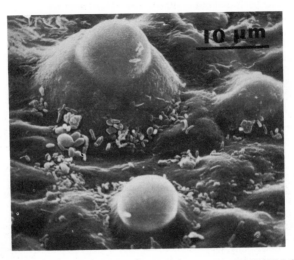

Fig. 4-7   Ocelli on the adaxial leaf surface of *Fittonia verschaffeltii.* SEM. Courtesy of Dr Jean Whatley, Botany School, Oxford University.

4.4.2   The reflection of light

Certain plants may, in extreme habitats, seek not to concentrate light but to reflect it. The soluble white secretion that forms in high light on the stems of *Kleinia articulata* from the Karroo desert may be as much concerned with reflecting light as with modifying the boundary layer and keeping water in. Some species of genera such as *Cotyledon, Aloe, Yucca* and *Sedum* may also use their thick crust of wax as a light reflecting surface.

A combination of light scattering and short-term water storage is achieved in the storage epidermal cells of *Halimione portulacoides* and, as we have seen, in *Mesembryanthemum crystallinum.* The stems and leaves of the latter appear studded with ice, hence its common name.

A dense tomentum of white hairs is a feature of many plants inhabiting dry hot regions, e.g. *Opuntia* species, *Convolvulus cneorum, Banksia stellata* and

Stachys lanata. Experiments show that removing these hairy coverings, as one might expect, dramatically increases transpiration, but their possible role in protective light scattering is less easy to measure. The peltate hairs of, for example, *Hippophae rhamnoides* have only limited light-scattering properties and are likely to be more concerned with water retention.

Living organisms other than plants use wax to reflect light from a surface. One of the tenebreonid beetles (*Cryptoglossa verrucosa*) not only grows wax filaments from miniature tubercles over the cuticle surface, which bear a striking resemblance to waxy plant surfaces, but can also change this surface depending upon the relative humidity. The so-called 'blue phase' beetles (with a blue/white iridesecence) are wax-rich and found in dry conditions. They can be artifically converted from a jet black colour to this iridescent phase by lowering the air moisture. When jet black they have little wax on their cuticles and this is their normal appearance in high humidity.

The cochineal beetle (*Dactylopius coccus*) (Fig. 4-8) likewise has a waxy epicuticle similar to that of many plants, correlated both with the need to reduce water loss and to defend itself against insects and birds. The waxy powder readily breaks off and clogs mouth parts and feet.

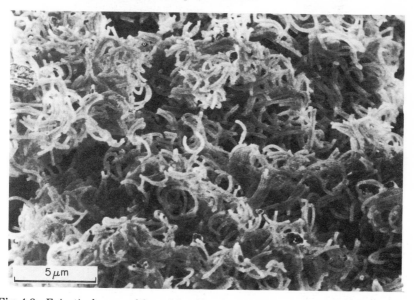

Fig. 4-8   Epicuticular wax of the cochineal beetle (*Dactylopius coccus*). SEM. Courtesy of Dr F. Baranyovits, Department of Agriculture and Horticulture, Reading University.

4.4.3   Light reflection and surface structure in petals

Petals are conspicuously unwettable, an absolute requirement of their dependence upon insect vision and the absorption and re-radiation of specific bands of the spectrum. However they rarely use the conventional light-scattering and water-repelling features of leaves, e.g. epicuticular particulate wax or trichomes, but rely extensively on epidermal cell shape (Kay, Daoud and Stirton, 1981).

In the above study of over 200 species from sixty angiosperm families six basic types of petal epidermal anatomy were found: papillate (Fig. 4-9), multipapillate (with in addition surface striations), reversed-papillate, multi-reversed papillate, lenticular and flat. Of which by far the greatest number, 112 species, are of the simple papillate type. The pigments such as anthocyanins and ultraviolet-absorbing flavonoids are usually confined to the vacuoles of the epidermal cells and light is usually reflected back from an aerenchymatous unpigmented reflective mesophyll. In certain *Ranunculus* and *Anemone* species however the light glances back from a specialized layer of starch grains.

Fig. 4-9 Light-scattering papillae on the adaxial petal surface of a daffodil (*Narcissus pseudonarcissus*). Fixed and critical-point dried. SEM. Courtesy of Mr D. Kerr, Botany School, Oxford University.

4.5   Thigmotropic Responses

Epidermal cells can also receive, transduce and transmit external signals other than light. Many plant organs respond to touch, demonstrating both thigmotropism (i.e. directional) and thigmonasticism (non-directional). Stamens will push up against insect pollinators as in *Portulaca grandiflora* and the pollen tubes of *Lilium longiflorum* will orientate their direction of growth with respect to the axis of an inert mesh.

Haberlandt (1914) believed that specialized epidermal papillae on stamens, the 'tactile' papillae, were the site of reception of this stimulus. But more recent work has shown that it is not necessary to touch these specific papillae for the stamen reaction to take place. Moreover, on the thigmonastic pea tendril and the pollen tubes no such papillae are present at all. Apart from the fact that the motor cells appear to respond by major turgor changes very little is known about the epidermal reception, transduction or delivery of the significant message.

4.6　The Wetting of Leaves and Surfaces by Water

4.6.1　Water repellency

Plant surfaces range through the complete spectrum from being highly water repellent to actively absorbing liquid water or water vapour. Most spores and pollen are highly hydrophobic which may be a direct aid to their dispersal (see Chapter 6); seeds are commonly hygroscopic (section 6.3.2) whereas most leaves, with some exceptions, are relatively unwettable on the upper surface and wettable below.

A convenient measure of the wettability of a leaf surface is the contact angle. Figure 4-10 shows a diagrammatic representation of static angles. Real life situations, where droplets fall or are projected onto plant surfaces and may rapidly dry, are better represented by dynamic measurements, e.g. the angles formed by advancing droplets rolling across a surface or drying droplets as the volume of a drop falls. However such dynamic angles, although real, are difficult to measure.

Static contact of plane surfaces of plane waxes are about 107°–108° for alkanes, 103°–105° for esters, ketones and secondary alcohols ranging down to

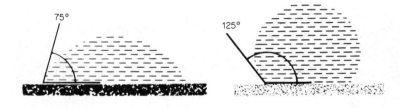

(a)　Wax-free plant cuticle. Small contact angle;
water drop covers a large surface area

(b)　Smooth wax surface. Larger contact
angle; droplet covers a smaller area

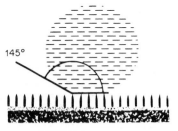

(c)　Crystalline wax surface, or ornamented
cuticle.
Largest contact angle. Droplet is in contact
with tips of the projections only

Fig. 4-10　Contact angles of water droplets with plant surfaces.

triterpenoids with angles of 89°–95°. Exotic substances such as PTFE (polytetrafluoroethylene) have angles only as high as 108°. However, many leaf surfaces have contact angles up to 150° (Holloway, 1968). These are not true surfaces, but the droplet is held away from the surface by large numbers of acicular projections of wax (Figs 2-2, 3-2 and 4.10C). A 50 μm droplet would make contact, on such a surface as Fig. 3-2, with over 1000 separate points.

Structures other than surface wax, plane or acicular, can produce high contact angles. Closed patterns of trichomes (Figs. 3-7 and 4-4) and the ridges on many grass leaves can create high contact angles unaffected by chloroform washing.

4.6.2   Water absorption

Some plants, like the epiphytic genus *Dischidia* (Fig. 8-5) from S.E. Asia have developed pitcher-like leaves which catch rain, stem drip, insect frass and decayed plant material. *Dischidia* has no conventional root system to absorb water and mineral salts, but adventitious roots grow into the pitchers and presumably absorb directly from what they accidentally trap. Unlike the superficially similar carnivorous pitcher plants, the inside of the pitcher does not appear to absorb anything directly through the cuticle. The surface is waxy and the stomata like Fig. 3-6 are sunk in deep crypts which prevent them from being flooded by water.

Plants may physically intercept and concentrate water from diffuse sources such as fog. Philips in 1928 exposed two rain gauges under similar conditions. One was unmodified. Over the other was suspended a screen of woven conifer branches with needles. The screened gauge collected almost twice as much precipitation as the control in a single season.

The paired needles of *Pinus sylvestris* are heavily waxed and cutinized and virtually impermeable to water in both directions. But rain and mist accumulate in the narrow gap between the paired needles, coalesce into small drops and run down into the needle base within the enclosing sheath. Here the cuticle is thinner, the wax absent and the water readily absorbed.

The whole maize plant acts in a similar way. The leaves, up to the fifth leaf, are waxy and unwettable, as is the stem. Light rainfall, which under normal conditions is lost by evaporation, is concentrated by the posture of the leaves into a significant stem drainage and exploited by roots developed close to the stem.

Prosopis tamarugo, a form of mesquite, grows in the extremely arid Atacama desert of Northern Chile. However nocturnal humidities there occasionally reach 80–100%. The leaves may fall slightly below ambient temperature at night, because of the outgoing long-wave radiation and a visible dew may form on the foliage. *P. tamarugo* has two types of epicuticular wax on its adaxial surfaces, numerous small plates 0.2–0.8 μm across and a smaller number of larger plates up to ten times as large, in profile like 'moose antlers'. The closely-related *P. velutina* normally forms only the small plates and does not normally grow in a dew-forming environment. However under artificial conditions where dew was induced to form on the *P. velutina* leaves, but less effectively than on *P. tamarugo*, it too formed, on the young developing leaves, the large wax plates like those on *P. tamarugo*. It is suggested that these projecting spikes are a development to promote a copious dew formation which may improve the plant's water balance and potential for survival under xeric conditions (Hull *et al.*, 1979).

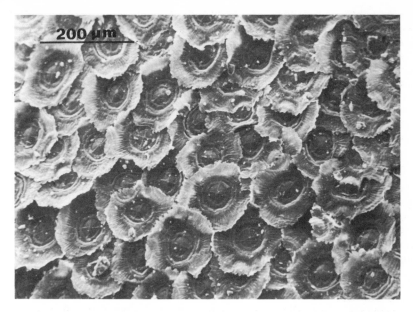

Fig. 4-11   The water-absorbing scales on the abaxial surface of a leaf of *Tillandsia duvalliana*. SEM. Courtesy of Mr D. Kerr, Oxford.

Many species of plants such as *Diplotaxis harra*, *Heliotropium luteum*, *Convolvulus cneorum*, some rhododendrons and many bromeliads (Fig. 4-11) have some form of water-absorbing hairs usually on the lower side of the leaf. These hairs take many forms, but basically expose some water absorbing surface when the air is wet. The water can be transmitted more or less irreversibly into the body of the leaf. When dry the scale or trichome commonly contracts reducing the possible transpiration area. In a *Tillandesia* these water absorbing structures occur all over the surface and water absorption occurs only when the surfaces are wet with rain or dew, but in the species which form leaf rosettes, the scales are distributed within the overlapping leaf bases, where water collects after rain, and the uptake is distributed over longer periods of time. In the carnivorous *Drosophyllum lusitanicum* (Fig. 5-5) which grows only down the Atlantic coast of Spain and Portugal, there are no water absorbing glands or scales, nor a substantial root system or storage. In common with most other carnivorous plants the cuticle is thin over much of the plant and yet the whole stem survives, catching insects on its mucilage coated glands through the whole of the very hot summer. It may use each sugary gland head as a osmometer, absorbing water from the mists that form each morning where the cold Atlantic meets the hot Iberian coast. Elleman and Entwistle (1982) (and see section 5.2.3) speculate that the basic cations secreted onto cotton leaves may have a comparable role and the desert-living *Nolana mollis* (Mooney *et al.*, 1980) may also absorb water with a similar hygroscopic crust.

4.6.3   Hydropoten

Literally 'water-drinkers', are multicellular epidermal structures, which apparently serve for the absorption of water and mineral salts. They are found mainly in the abaxial epidermis of the floating leaves of dicotyledons, particularly the young, unfolding leaves of such genera as *Cabomba*, *Nymphaea*, *Nuphar* and *Caltha*.

The general pattern of development seems to be that these structures usually begin life as mucilage-secreting glands. These abscise early; apparently then becoming permeable to a range of water-soluble dyes such as neutral red and methylene blue and take on a water- and salts-absorbing role.

4.7   Surfaces and Frost Protection

When leaves of frost-susceptible species are examined after they have been damaged by frost the intercellular spaces are often filled with water. These injected areas often turn necrotic as they thaw. Leaves that are wet just before the onset of the frost in general seem to be those most severely affected. There would seem to be two ways in which frost-hardy plants may protect themselves from such damage. The most obvious is to shed free water from the leaf surface by some hydrophobic system and the second is by adding anti-freeze to the liquid if it remains in position.

Eucalyptus delegatensis seedlings kept in high humidity are readily killed by simulated frost and those in shade are more susceptible than those in full sunlight. There is some evidence that glaucous ecotypes of this and at least eight other species of *Eucalyptus* may benefit by being able to survive in frosty areas.

In afro-alpine plants such as *Lobelia keniensis* water is held in imbricate leaf bases. This water is actively secreted and persists in dry weather. Water soluble materials discharged along with the water prevent freezing even in the coldest weather and the young central part of the rosette is thus protected. Young root tip surfaces, with their coating of mucilage (see section 7.1), may to some extent be preserved from frost in the same way.

Thus we can see that, although the leaf is constrained by its rigorous requirements as a solar energy collector, it nevertheless retains flexibility in its response to the environment. Light remains the dominant pressure in all habitats, but may be absorbed or scattered, re-radiated at different wavelengths, used as a source of orientation of the leaf or even guidance to an insect. The cuticle, although usually considered little more than the glass in a greenhouse may nevertheless have passive roles in a broad spectrum from frost resistance through drought and mechanical damage protection to gas exchange. Its more positive, aggressive roles we shall observe in the next chapter.

5 Plant Surfaces in Defence and Attack

5.1 In Defence

5.1.1 Wind damage and weathering of plant surfaces

Even gentle breezes can damage plant surfaces, buffeting leaves and brushing them together. Continuous touching and abrasion smooths and flattens the wax, especially on the crests and ridges of the epidermal cells. Eventually the cuticle and underlying cell walls also become damaged. Tips and edges of leaves are especially vulnerable and in turn cause local damage where they repeatedly hit other leaves. Stronger winds cause leaves to strike each other with sufficient violence to break trichomes and disrupt the cuticle, the epidermal cells and the underlying mesophyll cells, causing lesions into which pathogens can enter and from which tearing can start (Carlquist, 1956; Lipetz, 1970). Even the frictional force of the airstream over a leaf may be sufficient to smooth delicate wax structures. Certainly wind-carried dust, sand and soil particles, rain and hail damage not only the wax and cuticle, but also cause pinpoint lesions where cells are killed at the point of impact.

Young leaves have some capacity to regenerate wax, but after maturity the fine structure is progressively degraded by weathering. Waxes differ in their resistance to abrasion. In *Eucalyptus*, there are two distinct types. The tubular β-diketone-rich waxes are readily rubbed off, while plate type primary alcohol-rich waxes are more resistant (Hallam and Juniper, 1971).

Leaves and fruits from which wax is abraded or dissolved transpire faster than those with intact wax. This is exploited commercially in the fruit-drying industry, e.g. raisins and currants, where rapid drying has economic importance. Since the fine structure of wax determines wettability (see section 4.6), weathering enhances the retention of water films, encourages greater leaching losses and improves both the spore-trapping efficiency of surfaces and the germination prospects of trapped spores. Lesions can be ideal sites for pathogen invasion (see later) especially if the pathogen is already present, but wounds are rapidly sealed by deposits of tannins and other antibiotic metabolites.

Recent studies have shown that grasses and sycamores (*Acer pseudoplatanus*) may lose 10 to 50% of the leaf area in a season due to wind induced lesions and tearing. There is a reduction in photosynthesis in the short term, but plants seem to compensate for this loss by increasing rates of photosynthesis in the undamaged parts of the leaf.

The relationships between wind damage and crop yields are complex and poorly understood (Monteith, 1973). Shelter trials suggest that up to 60%

improvement in yield is possible in sheltered crops, especially where yield is not already limited by drought or poor nutrition. This is probably due to improved water economy and photosynthesis. However, sheltered crops transpire from a larger leaf area and the water contents of soils are not always improved by shelter. Shelter also reduces turbulent mixing over protected canopies. This may deplete carbon dioxide concentrations in the atmosphere surrounding actively photosynthesizing plants and reduce the efficiency of photosynthesis when measured on the basis of equal leaf areas, despite the increase in overall production.

5.1.2   *Effects of pollution on plant surfaces*

Stands of vegetation usually have more surface area than the ground they occupy and thus efficiently intercept atmospheric constituents. This characteristic, essential for efficient photosynthesis, also makes plants vulnerable to the damaging effects of atmospheric pollution.

Atmospheric pollutants arrive at plant surfaces as particulates (dry dusts, liquid droplets, semi-solid particles) (Fig. 8-2), in solution or suspension in rainfall (soluble gases and particulates), or as gases.

Dusts and particulates can be chemically inert substances such as soil particles, silica, mica and soot (Fig. 5-1), or are more or less chemically reactive, e.g. cement dust, fertilizers, ammonium sulphate, sulphuric acid, and metal oxides. Almost all of these modify the wettability, permeability, reflectivity, water loss, gas exchange and other properties of plants. Acidic aerosols of sulphuric and nitric acids produce small necrotic areas on leaves and may cause

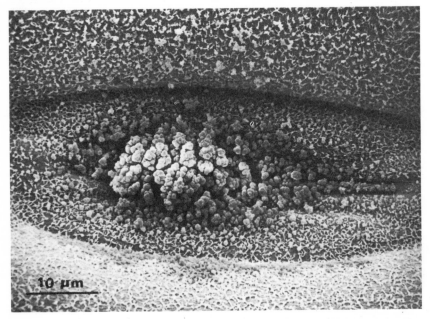

Fig. 5-1   The adaxial surface of a wheat leaf (*Triticum vulgare*) showing soot particles (from an oil-fired boiler house) clogging the aperture of a stoma. SEM. Courtesy of Mr D. Kerr, Oxford.

sublethal toxic effects on surrounding cells. Cement dust is strongly alkaline and capable of hydrolysing the cuticle, thereby increasing its permeability. Opaque particles reduce light available for photosynthesis. Particulates can also influence the epiphytic flora both by changing the pH of the surface and by toxic effects and they sometimes provide sources of nutrients (see section 8.2 and Fig. 8-2). Although it is conjectured that pollution by particulates may alter crop yield the economic consequences have not been estimated.

The most important gaseous pollutants affecting economic crops are sulphur dioxide, ozone, nitrogen oxides, fluorides, peroxyacyl nitrate (PAN) and hydrocarbons. Plants readily absorb soluble gases such as sulphur dioxide and nitrogen dioxide directly via the stomata, but much is absorbed at the plant surface especially when the surface is damp. Sulphite ions directly inhibit fungal activity, germination and infectivity of pathogen spores, with possible beneficial effects (see below). On the other hand the general weakening of plants by exposure to pollutants may predispose them to pathogen attack. It is not known whether the inhibition is due primarily to pH changes or toxicity, since the two are difficult to distinguish experimentally. Much depends on the concentration however. Fungi differ in their sensitivity to sulphur pollution and for each species, a sub-inhibitory concentration may actually stimulate growth by providing a source of nutrient sulphur. Many pathogens are sensitive to the sulphur dioxide concentrations experienced in the vicinity of towns. Tar spot of sycamores (*Rhytisma acerinum*) and black spot of roses (*Diplocarpon rosae*) scarcely occur in towns although they are ubiquitous elsewhere, making them useful negative indicators of atmospheric pollution.

High concentrations of gaseous pollutants cause visible or 'acute' damage to plants. Sulphur dioxide causes brownish lesions in the interveinal areas of leaves, while ozone-damaged leaves appear bronzed. Pollution-damaged plants often appear chlorotic or prematurely senescent. In western Europe gaseous pollutant concentrations high enough to cause visible damage are very local, usually confined to urban areas. Subacute concentrations, typical of agricultural areas, have subtle effects on crop physiology and biochemistry which may affect yields in the long term. The damage caused by pollution may be minimal when nutrient deficiencies limit yields and sulphur dioxide and sulphates intercepted by leaves may be a valuable source of nutrient sulphur, especially where the stimulation of growth by nitrogen fertilizers creates an increased demand for sulphur.

5.2   The Assault of Pathogens on the Leaf Surface

5.2.1   Fungal deposition and penetration

The physical and chemical nature of the plant surface may influence the position of landing and the site of penetration by a pathogen (Martin and Juniper, 1970; Dickinson and Preece, 1976; Blakeman, 1981). A water-borne fungal spore will find it difficult to land on a water-repellent surface such as Figure 3-2. On the other hand a 'sticky' spore will have no difficulty in landing on almost any form of surface. Once landed most spores will not germinate unless immersed or in a very moist atmosphere. At 80% humidity *Plasmopora viticola* will infect mature vine leaves, but at 70% will only attack young vines. The extent to which a plant surface can shed rain, or conversely coacervate dew drops will obviously affect the success of pathogens on its surface.

Pollen grains, with their soluble surface proteins (Fig. 6-1), can stimulate the germination of fungal spores. When fruits of the strawberry (*Fragaria ananassa*) are sprayed with conidia of *Botrytis cinerea* no infection occurs if the anthers are removed, but a rapid rotting develops if the anthers are present. Pure pollen and pollen extracts promote the germination of the conidia on strawberry petals and broad bean (*Vicia faba*) leaves. Wheat anthers likewise promote the infection of the wheat spikelet by *Fusarium gramineum*. The spikelets are resistant prior to anthesis.

Fungi often secrete specific antibiotics which inhibit bacteria and other fungi. When spores of saprophytic fungi outnumber rust uredospores the germination of the rust is inhibited and the non-parasitic fungi *Aureobasidium pullulans* and *Dendrophoma obscurans* both secrete substances which suppress the pathogen *Botrytis cinerea*.

Indirect antagonistic effects can also occur, resulting from the stimulation of the hosts' defence mechanism. The presence of a pathogen on the plant surface may induce phytoalexin production by the host and other potential pathogens may also be inhibited. This may explain why *Uromyces phaseoli*, a pathogen of beans (*Phaseolus vulgaris*), will prevent the infection of sunflowers (*Helianthus annuus*) by *Puccinia helianthi* despite being unable itself to attack sunflowers. On the other hand, the weakening of the host tissue by a pathogen can enable a secondary infection to occur by a weak pathogen which would not initiate an attack on healthy tissue. For example, the soil fungus *Fusarium roseum* can invade leaves of snap-dragon (*Antirrhinum* sp.) via the open pustules of *Puccinia antirrhini*, although it is rarely the primary pathogen.

Various influences may orientate the hyphal tube once growth has started. The germ tubes of *Puccinia* can detect and grow across ridges in a membrane and may be able to orientate on a leaf surface as they 'search' for stomata. Many pathogenic fungi search for open stomata, e.g. downy mildew (*Sphaerotheca humuli*) on hops (*Humulus lupulus*) and *Puccinia* spp. on a wide range of hosts. Others such as *Phytophthora infestans* on the potato (*Solanum tuberosum*) will pass through the stomata or penetrate the cuticle. Most cuticle borers, after a brief period of mycelial growth, will form an anchorage called an appressorium, from which an infection thread forces through the wax, if present, and the cuticle. The evidence for widespread cutinase in hyphal tips may be disputed, but there is no doubt that the fungal ability to dissolve waxes is very common. Ridges and grooves are frequently seen on epicuticular wax surfaces colonized by fungal pathogens. Fungal tips are able to exploit small cracks or defects in a cuticle. The thicker cuticle of the mature leaves of *Populus tremuloides* seem to be physically resistant to penetration by *Colletotrichum gloeosporoides* whereas the young leaves are readily infected. However, this resistance need not solely be due to extra cutin. The ornamental shrub *Euonymus japonicus* is severely attacked in southern England by the mildew *Oidium euonymijaponicae* although it has a very heavy cuticular layer on both surfaces. Moreover, some fungi, such as the ascomycete genus *Vizella*, spend the greater part of their life cycle in the thick cuticle of their host (see Chapter 8-2) (Swart, 1972). It is difficult, therefore, to believe that a cutinase may not sometimes be a feature of the fungal armoury.

5.2.2   *Bacterial colonization and penetration*

Bacteria on plant surfaces are even more dependent on the humidity of the phyllosphere than fungi. Fire blight (*Erwinia amylovora*) of apples and pears will not grow at 97% humidity, grows slowly at 98% and satisfactorily at 99–100%. Rain and dew are particularly important for bacterial infections and it is not surprising that many bacteria favour a stomatal entry, where the humidity is naturally high. *Erwinia* is probably carried by bees and wasps and commonly infects the nectaries of pears. The bacteria lodge in the nectar secretion and as the exudate dries may be drawn into the sensitive nectary chamber. Other bacterial infections can take place through small wounds in the cuticle and *Agrobacterium tumefaciens* can penetrate tomato (*Lycopersicon esculentum*) leaves simply by being brushed across the surface with a smooth needle. The broken trichomes are apparently sufficient to provide ports of entry. Thus the normal abrasion of leaf on leaf is adequate damage for many bacteria to exploit.

Bacteria multiply rapidly in drops of water on a leaf surface. Ultra-violet irradiation of water droplets on *Beta vulgaris* and *Chrysanthemum* sp. failed to decrease the bacterial count, whereas similar populations on glass slides were eliminated. The bacteria are probably sheltered from UV radiation by folds or cracks in the surface of the cuticle, are obviously indifferent to the effects of sunlight and this probably indicates that bacteria are part of a normal phylloplane flora.

5.2.3   Virus adhesion and penetration

Plant viruses have no moving parts and can only be assisted through the plant surface by a range of vectors. Almost anything can thrust a virus particle through a plant surface, from the fungus *Olpidium brassicae* carrying a virus through the softer root surfaces into the roots of cabbages, to the hands of a gardener working amongst tomatoes and spreading, through contact, tobacco mosaic virus (TMV). Biting and sucking insects, aphids and mites transfer viruses on or in their proboscies and punch these through the cuticle as they seek the sugar-rich sieve tubes beneath. Nematodes too are common vectors for viruses and most are pushed through the more succulent regions of the root. Viruses, which are more than an order of magnitude smaller than bacteria, lodge in almost any leaf crevice and are very difficult to remove. Cabbage (*Brassica*) leaves were experimentally coated with the granulosis virus of the cabbage white butterfly (*Pieris brassicae*) to initiate a non-toxic mechanism of biological control. The virus was not removed to any significant extent by 5 hr of simulated rain or by scrubbing in detergent and rinsing in water. Under natural conditions of weathering the virus persisted on the surface for up to 4 months. Other viruses, e.g. *Smithiavirus pityocampae* on *Pinus nigra* adhere just as well. TMV, however, can readily be washed from the leaves of *Nicotiana tabacum* and *Datura stramonium*.

It might be supposed that, once firmly adhering to a plant surface, a virus would remain viable and immune to degradation other than that resulting from UV light. This may not necessarily be the case. Elleman and Entwistle (1982), amongst other workers, noticed that nuclear polyhedrosis virus on the surfaces of cotton (*Gossypium hirsutum*) plants was rapidly inactivated both in laboratory and field conditions. They attributed this in part to the high pH (between 7.4 and over 10.0) on these surfaces, deriving from the secretion by the surface glands of the basic cations magnesium, calcium and potassium, which are probably present as carbonates and bicarbonates.

The prospect of entry for a virus either by abrasion, insect damage or handling must be enhanced if it can attach to a leaf surface for a considerable period of time and stay viable. It has been noticed that if a trichome is broken while still alive a small drop of cytoplasm is exuded and then contracts back. As we have seen in the infection by the bacterium *Erwinia* such a withdrawal could readily pull viruses back close to living cells.

It was shown in 1953 (Tinsley) that *Nicotiana glutinosa* and *N. tabacum* are both more susceptible to artificial virus infection if well watered and also if kept in the shade. The effects may be separate, but are additive and may result from a thinner cuticle and larger and looser cells in the epidermis, spongy parenchyma and palisade layer.

5.3   The Surface's Defence against Insects

5.3.1   Defence by chemical means

Plant surfaces may repel insects by chemicals. The leaves and flower buds of *Rubus phoenicolasius* are unattractive to the raspberry beetle (*Byturus tomentosus*); the plant is heavily waxed and the wax is richer in acidic material than other *Rubus* waxes. Likewise, the waxy sprouting broccoli (*Brassica oleracea* var *italica*) is more resistant to attacks of the flea beetle (*Phyllotreta albionica*) than the glossy-leaved wax-free mutant. However, against the usual trend, the normal waxy plants of marrow-stem kale (*B. oleracea* var *acephala*) have large colonies of the cabbage aphid (*Brevicoryne brassicae*); the non-waxy mutant normally has none.

Glandular trichomes secrete an immense range of compounds such as α-pinene, menthol, cineole, menthone, sabinene, camphor, limonene, the phenolics, the flavones and pirimetin. Rarer compounds include cinecine from *Cnicus benedictus* and absinthe from *Artemisia absinthum*, in each case responsible for their bitter taste. Some of these compounds are proven toxins. The alkaloids from the trichomes of *Nicotiana*, for example, are toxic to green peach aphids. The resistant species of tobacco produce nicotine, nornicotine and anabasine, whereas *N. megalosiphon*, which has a rather lower resistance, only produces nicotine. Sometimes the secretions may poison the insects, sometimes, as we shall see later, they are physically trapped in a gummy exudate.

5.3.2   Defence by physical means

The potato leaf hopper (*Empoasca fabae*) is one of the most abundant phytophagous insects found on soybean (*Glycine max*) and causes extensive damage to this and other legumes. Depending on the genotype, soybean leaves range from almost glabrous to densely hairy. In test plots the glabrous forms may have up to 15 times as many leaf hoppers per unit leaf area. Another study showed a similar correlation between hairiness and resistance to the leaf hopper in cotton (*Gossypium hirsutum*). However, hairiness lowered cotton's resistance to the fleahopper (*Psallus seriatus*).

Hooked hairs such as those on *Phaseolus* leaves have traditionally been used to control bed bugs. The leaves were scattered about the bed to trap the roving predators. *Empoasca* and *Aphis fabae* can be caught in the same way on the same leaves.

5.3.3 Defence by mimicry

The larvae of *Heliconius* butterflies are voracious predators of *Passiflora* species in the Americas. But *Passiflora cyanea* forms 'dummy' eggs on its leaves of the right size, shape and colour, which apparently convince *Heliconius* females that a particular leaf has already been colonized. The females leave such 'infested' surfaces alone and lay their eggs elsewhere (Williams and Gilbert, 1981).

5.4 Plant Surfaces in Attack

5.4.1 Secretion and the modification of chemicals on plant surfaces

When chemicals fall on plant surfaces they rarely remain unmodified by their new environment (see also section 8.2). Apple (*Malus*) and dwarf bean (*Phaseolus vulgaris*) leaves, if coated with dried copper sulphate, can dissolve the copper and thereby enhance its fungistatic properties. In a similar way a dithiocarbamate fungicide (*Zineb*) is affected by substances secreted from a wide range of leaf surfaces. In most cases, e.g. grapes (*Vitis*) the fungistatic action of the chemical is enhanced.

5.4.2 Surfaces that just kill

Insects predate plants and in turn a wide range of insects die or are killed on plant surfaces. Whether the plant makes any use of the corpse varies from plant to plant. A trichome secretion on some hybrids of *Solanum tuberosum* and *S. berthaultii* can catch and glue down aphids moving over the leaf surface (Fig. 5-2). The European catchflies (*Silene* sp.) are commonly crusted with insects trapped on their viscid hairs.

The seeds of *Capsella bursa-pastoris* are heavily coated with mucilage, and mosquito larvae and nematodes are often found glued to them. There is, however, no evidence of any advantage to the seed from this fly-paper device (Barber, 1977).

Roridula from South Africa kills a wide range of insects on its leaf surfaces by glueing them down on its vascid surface. Both Charles Darwin and for a while Francis Lloyd considered *Roridula* a true carnivorous plant, but the insect is not degraded *in situ*, the leaf finally falls to the ground carrying with it a burden of dying and decaying flies. Whether this defends the plant from predation is obscure. Carlquist (1976) suggests that the rain of protein-rich leaves to the ground immediately around the plant is of nutritive value – a sort of instant, high-nutrient compost. Several *Solanum* species as we have seen above kill aphids and this may assist them indirectly in slowing the spread of virus diseases. Several species of *Passiflora* kill heliconiine caterpillars on their leaf surfaces by spiking them with their short pointed leaf trichomes. Again this would seem to be a direct attempt to avoid predation by the caterpillar. In none of these cases is use made directly of any of the insect corpses, although the plant in some instances may gain indirect advantage.

The larvae of all species of cattle ticks (Ixodidae) climb plants to enable them to transfer to a passing host. Contacts with the hosts are infrequent so the ticks must often wait in the pasture for several weeks. Where the ticks climb sticky stems such as those of molasses grass (*Melinis minutiflora*) they may become immobilized and slowly die. The effect however is protracted and does not appear to be of any economic significance.

Fig. 5-2   An aphid (*Myzus persicae*) glued to the glandular exudate on a leaf surface of *Solanum polyadenium*. SEM. Courtesy of Dr R. Graham, Rothamsted Experimental Station, Harpenden, Herts.

However, in some species of the highly productive and nutritious tropical legume *Stylosanthes* there are glandular trichomes which secrete both a viscous fluid and some sort of toxic vapour. Varieties of *Stylosanthes scabra* and *S. viscosa*, which are native to S. America but are widely planted in Australia, produce sticky secretions which immediately trap ascending larvae of the cattle tick *Boophilus microplus* (Sutherst, Jones and Schnitzerling; 1982, Fig. 5-3). Some toxic component, which appears to be effective only in the vapour phase, rapidly kills the entrapped larvae. The closely related *S. hamata*, whilst effective in trapping the larvae, does not apparently kill them. The *Stylosanthes* plants do not here appear to benefit from the insects trapped and killed on their surfaces, but it must be noted that they are transplanted from their normal habitat. On the other hand the economic possibilities of a nutritious forage plant which also controls a major cattle pest are considerable.

If it is difficult sometimes to see the selective advantage of this indiscriminate insect killing perhaps we should look again at the very early suggestions of Kerner (1878). Many of these insect entangling devices he interpreted as mechanisms for inhibiting the predation of flowers and their nectaries, pollen, fruit or seeds by terrestrial insects. The relatively intelligent and taxonomically consistent Hymenoptera were obviously to be encouraged, but this also invited the unwelcome and indiscriminate terrestrial plunderers. From such purely defensive devices, he speculated, some carnivorous systems, e.g. that of *Pinguicula*, might have evolved.

Fig. 5-3 Cattle ticks (*Boophilus micropus*) trapped and killed on the adhesive and toxic glandular secretions of the legume *Stylosanthes scabra*. The sticky glandular hairs are a little over 1 mm long. SEM. Courtesy of Dr R. W. Sutherst, CSIRO Division of Entomology, Indooroopilly, Queensland, Australia.

5.4.3 Surfaces that kill to eat: attracting the prey

Flowers seek to attract their usually specific and highly sophisticated pollinating insects by colour guides (concentrating on the blue end of the spectrum) and seductive odours, mostly from nectaries. Carnivorous plants are more catholic in their choice of prey, but rarely choose to interest Hymenoptera since they are mostly too powerful for their mechanisms.

It has long been known that freshly hatched greenbottles, blowflies and fleshflies all head immediately for glistening objects. A few plants, e.g. *Parnassia*, where the sterile stamens form gleaming knobs, may use this response as a pollination mechanism, but the greatest advantage of this phenomenon seems to have been taken by a range of carnivorous plants. The glands of *Drosophyllum* (Fig. 5-5), *Drosera* and *Pinguicula* (Fig. 3-8), with their long-stalked trichomes topped with iridescent mucilage, are designed to catch the sun (and insects). Surface nectaries create a ring of specular globules around the rim of the pitchers of *Nepenthes* and *Sarracenia* and fringe the edge of the leaf trap in *Dionaea*. Their odours, if they produce any, seem to have little attraction for bees, but their attraction for a whole range of flies is manifest (Darwin, 1875; Lloyd, 1942).

5.4.4   Surfaces that kill to eat: absorbing the constituents

True carnivorous plants may be unsophisticated in every other detail, but possess the mechanism of absorption. *Darlingtonia californica* (Sarraceniaceae) is a simple pitcher trap into which flies and other arthropods fall. They are hindered in their escape by a system of downward-pointing hairs inside the pitcher and the generally viscous nature of the fluid inside. However, although these modified leaves are capable of absorbing the remains of the decaying insect there is no evidence of the active secretion of proteolytic enzymes; this is left to bacteria present in the trap or those brought in on the prey's own surfaces.

On the other hand *Dionaea* secretes a wide range of proteolytic enzymes, a chitinase and water from its surface glands (Fig. 5-4) and absorbs the degraded insect components through the same gland. *Drosophyllum*'s gland (Fig. 5-5) like that of *Drosera* also has a multiple function secreting mucilage and, on stimulus, proteolytic enzymes. But only in *Drosera* can the same gland absorb proteins and other materials (Darwin found *Drosera* responded well to milk) and at least in the peripheral glands the stalk has both a tropic and nastic power of movement in response to an indirect chemical stimulus.

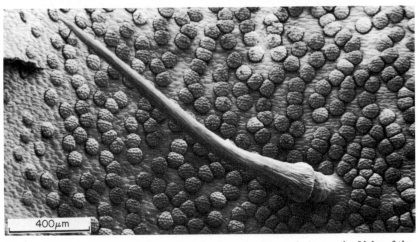

400μm

Fig. 5-4   The secretory and absorptive glands and a trigger hair on a leaf lobe of the Venus Fly-Trap (*Dionaea muscipula*). Each gland is about 80 μm across and trigger hair about 1.3 mm long. SEM. Courtesy of Mr D Kerr and Mr P. Rea, Oxford.

5.4.5   Surfaces that kill to eat: restraining the prey

An insect placed in the middle of a *Drosera* leaf will cause the peripheral glands to bend inwards. This is an irreversible growth process and can normally only be performed three times. The speed is impressive, 270° of arc in 1.5 mins, and the effect is to trap a struggling insect in a greater mass of mucilage. Thus

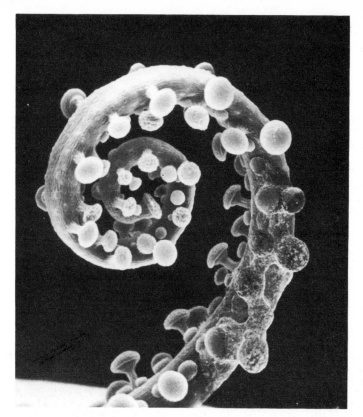

Fig. 5-5   A crozier tip of a leaf of *Drosophyllum lusitanicum* showing both stalked secretory glands (S) and sessile absorptive glands (A). As the secretory glands mature they begin to produce the trapping mucilage (M). This whole crozier tip is about 2 mm long. SEM. Courtesy of Mr. G. Wakley, Oxford.

the insect is drowned and a greater concentration of enzyme is brought to the potential food surface.

The surface mechanisms for restricting the escape of insects drawn into the traps of carnivorous plants are highly varied. Common to the genera *Sarracenia, Darlingtonia* and *Cephalotus* are the curved downward pointing hairs within the pitchers (Fig. 5-6) which combined with a surface coat of mucilage is difficult for most insects to climb. In the superficially similar pitcher plant *Nepenthes*, which is not closely related to the above genera, the general principle of alluring, trapping, killing and absorbing is much the same, but the downward pointing hairs within the pitcher are replaced by a system of small (c. 1μm) wax plates, secreted from the epidermal cells and overlapping like the tiles on a roof (Fig. 5-7). These detach easily, clog footpads, claws and mouthparts and render the escape of a wide range of arthropods virtually impossible.

Hairs may, as we have seen above, play a passive role in the entrapment of the insect prey. In *Dionaea*, however, and the closely-related *Aldrovanda*, touching

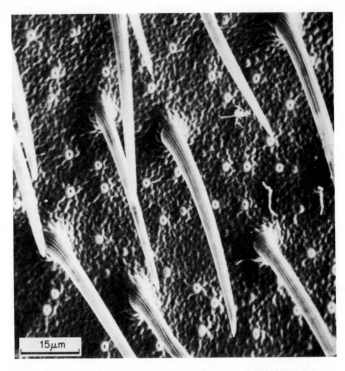

15μm

Fig. 5-6   The ribbed downward-pointing hairs inside the hood of *Sarracenia purpurea*. Each of these hairs is a little over 1 mm long and would, in the living plant, be coated with mucilage. SEM. Courtesy of Mr G. Wakley, Oxford.

the leaf-mounted triggers (Fig. 5-4) springs the trap into action, but in each case one stimulus is not normally enough and a second stimulus, following the first very closely, is necessary. By contrast, the triggers which activate the traps in all the genera of the Utriculariaceae (bladderworts) are instantaneous.

The movement of any of these trigger systems propagates an action potential across the whole leaf surface, but how this brings about the rapid closing of these trap mechanisms is still obscure.

The development of controllable vents, the stomata, in the 'greenhouse roof' was, as we saw in section 3.1 a very early possibly polyphyletic, but revolutionary, event in the evolution of the plant surface. From the simple stomata there may have developed the hydathodes and some other types of glands. A further development of the standard epidermal cell, but possibly a little later than stomata, were the various papillae and trichomes. In combination, these three structures provided the plant surface with all its potential for defence and attack and, in part, some of the sites for the exploitation of its chemical properties as we shall see in the next chapter.

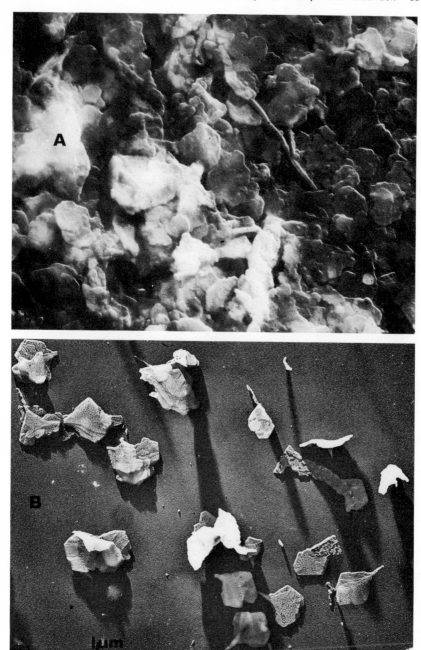

Fig. 5-7   The waxy scales on the inside surface of the pitcher of *Nepenthes* species. **A** The intact surface and **B** detached scales. Metal-shadowed carbon replica. TEM.

6 Plant Surfaces in Reproduction and Dispersal

6.1 The Surface in Pollen Retention, Recognition and Response

6.1.1 The stigmatic surface and the transmitting tissue

The surface of the stigma of a flower has a thin but perceptible cuticle and occasionally wax. Prior to pollination, as in *Petunia hybrida*, a sticky exudate containing sugars, amino acids and fatty acids oozes from the cells of the stigmatic epidermis, loosening the cuticle over the surface. This exudate presumably has a multiple function. It ruptures the cuticle above, provides a temporary skin prior to pollination and thus slows down water loss from a sensitive tissue. It is a mucilage on which the drifting pollen sticks, may nourish the pollen tube as it begins to grow and may contain recognition signals where incompatibility systems are operative (Heslop-Harrison, 1978).

Between the stigma and the ovary lies the transmitting tissue through which the pollen tube grows. In some species canals develop in this tissue prior to pollination and a thin cuticle can develop on the surface of these canals. But just prior to pollination this cuticle tends to disappear and the whole transmitting tissue softens and swells.

There is a waxy layer on the stigmatic papillae of *Brassica* species which can, without affecting the cutin underneath, be removed by chloroform. After a compatible pollination the pollen grain 'sticks' to the wax layer so that it is apparently pierced and the exine of the pollen touches the cuticle proper. If the pollination is incompatible the stigmatic papillae stay turgid, but if compatible the papillae collapse, the pollen tube penetrates the cuticle and grows down the style. There are, therefore, several phases of physico/chemical recognition: a sticking, a removal of the wax, an interconnection of the exine with the cuticle, the germination of the pollen and the penetration of the cuticle. It is now well established that germinating pollen grains of a number of species, not only of *Brassica* but also others such as *Diplotaxis tenuifolius*, possess both 'surface recognition' proteins (see below) and enzymes including cutinase. This cutinase is used to penetrate through the cuticle over the stigmatic surface.

Pollination in the gymnosperms is a little different. Here the appropriate pollen, light and windborne, is blown onto the micropyle where it is trapped on a drop of sugary secretion. This liquid is secreted from the ovule and, as it dries, the grains are drawn inward towards the nucellus which is reached with a short pollen tube.

6.1.2 The surface of the pollen grain

The surfaces of most pollen grains are coated with at least two types of proteins

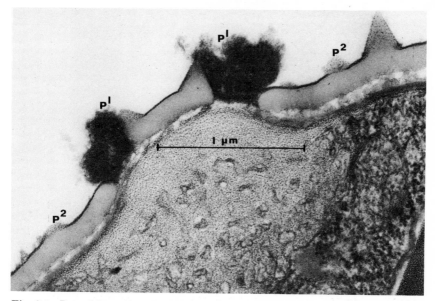

Fig. 6-1   Part of the surface of a pollen grain of *Urtica dubia*. Two types of proteins can be seen on the surface: (P^1) in the apertures of the pore and (P^2) surrounding the operculum. TEM. Courtesy of Dr E. Pacini, Siena.

(Fig. 6-1), soluble proteins which are washed away in a moment as the pollen comes in contact with a wet surface (and which cause hay fever), and those which adhere more tightly to the exine. The pollen grain will normally germinate if the 'recognition proteins' of which there may be several on an individual pollen surface, bring about correct mating between pollen and stigma. On receipt of the correct 'feed-back' the pollen tube will emerge (Fig. 6-2) and grow down the stylar canal.

6.2   Pollen Surfaces and Distribution

Pollen is commonly adapted to adhere to the body of the pollinating organism. A sticky oil, the 'pollenkitt' of some authors, sticks securely even to the chitin of glabrous insects. In many Orchidaceae the pollen remain clumped together in pollinia and, for example, *Platanthera* species place the sticky pollinia on the eyes or the proboscis of the pollinator, usually dipterans or hymenopterans. *Salvia pratensis* glues its viscid pollen to one spot on its visitor's back. The obverse of this mechanism is found where dry pollen is collected by a viscid stigma. Sometimes mucilage from the stigma is transferred to the insect vector which secondarily collects the appropriate pollen on the glue.

Wind pollination (anemophily) primary in the gymnosperms, is obviously a secondary evolutionary character in the angiosperms. In the angiosperms some families even show a transition between insect and wind pollination. *Urtica*, for example, is wind-pollinated but has relict nectaries. Wind pollen (usually 20–30 μm in diameter), is generally smaller than insect pollen, obviously a correlation

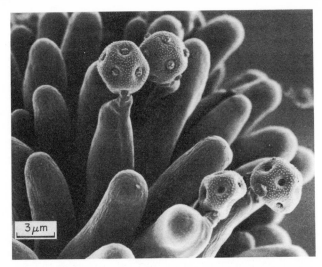

Fig. 6-2   Pollen of *Myosoton aquaticum* (Caryophyllaceae) on the stigmatic papillae. The pollen have germinated, penetrated the cuticle of the papillae and have begun to grow down the stylar canals. SEM. Courtesy of Mrs K. Novosel, Department of Botany, University of Edinburgh.

with air buoyancy. In the larger pollen of conifers, (Fig. 6-3) one or more air sacs give buoyancy without much increasing the weight. 150 μm pollen grains of the Pinaceae have been found 400 km from the source. In addition, some globular pollen grains become disc-like on drying, giving the better aerodynamic shape of a 'frisbee'. The grains of wind pollen are dry and do not adhere to one another.

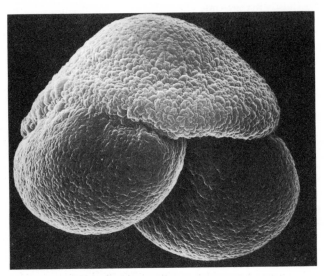

Fig. 6-3   The wind-borne pollen of *Pinus sylvestris*. SEM.

Typically insect pollen grains are both sticky and heavily ornamented and adhere both to the vector and to each other. The remains of small quantities of oil (pollenkitt) on the pollen surface of many wind species, e.g. *Plantago* spp. and grasses, hints at a former entomophilous existence. Conifers as primary anemophiles are probably free from pollenkitt.

The oily layer on some water plants enables the pollen, as in *Ruppia* and *Calitriche autumnalis*, to float up to the surface and drift onto stigmas exposed above water level. But other pollen, such as that of *Najas*, is wettable and slowly sinks down into the large trumpet-like submerged stigmas.

6.3 The Surfaces of Seeds

6.3.1 Dispersal mechanisms

The testas of most seeds consist of tough secondary thickened cells impregnated and waterproofed with tannins, and unaffected by humidity changes. However, some plants, e.g. the cranes' bills (*Erodium*), oats (*Avena*) and *Arrhenatherum elatius*, use changes in the water environment to bury seeds. *Erodium* and *Avena* seeds are pointed at one end with a beak at the other. From the pointed end hairs project backwards to serve as an anchor. When wet the median section of the beak is nearly straight and when dry twisted into a tight coil. Alternation of wet and dry periods causes the median section of the beak to coil and uncoil and, like a climber 'chimneying' up between crags, thrusts the seed progressively deeper into the soil. The depth reached corresponds to the point where diurnal fluctuations of humidity are no longer significant to the seed coat.

True seeds vary from those of some orchids 0.25 mm long and about 3 μg in weight up to *Lodoicea maldivica*, the coco-de-mer of the Seychelles, which can be 45 cm long and 3.0 kg in weight.

Acorn, chestnut and the above-mentioned *Lodoicea* are so massive they germinate where they fall, or are carried by animals. Birch (*Betula*) and thistledown (*Cirsium* spp.) are carried by the wind to earth but the seeds of orchids and bromeliads almost as light as pollen grains, can drift with the wind and lodge in the cracks and crevices of tree bark to start a new generation of epiphytes (Chapter 8). Perhaps the most striking epiphyte seed is that of *Chilochista lunifera*. Only 0.6 mm long it can blow into its new lodgement, but once wetted, the surface of the testa throws out 4 mm long helical threads which tangle and attach the seed to moist bark. Alder (*Alnus*) seeds are waterproof and unwettable and can float in water. Figs (*Ficus*) and tomatoes (*Lycopersicon*) can travel unharmed through the guts of animals and may indeed need to do so before they can germinate. Grasses and burdocks travel on the fur of animals. Some seeds are mucilaginous and stick to birds' beaks and claws. *Pisonia* of the Pacific, normally distributed by big birds such as terns and frigate-birds, may glue so tenaciously to others such as white-eyed warblers as to kill them. In others, e.g. the seeds of *Plantago lanceolata* the mucilage glues them to soil particles and inhibits their movement on the soil.

6.3.2 The surfaces of seeds in maintenance of viability and inhibition or enhancement of germination

The surface of a seed has almost as many problems to solve as that of a pollen grain, save only that of recognition. However, the seeds of a few species in the genera *Striga* and *Orobanche*, and possibly some other root parasites, recognize substances exuded from roots. In the absence of these as yet unidentified stimulants they will not germinate.

The seeds of *Blepharis persica* have unicellular hairs on the testa. On wet ground these hairs swell, raise the seed from one end to an angle of 30–45° to the soil surface and thus bring the micropyle into contact with the wet soil surface. This process only lasts a few minutes; the tips of the hairs then become mucilaginous and dry and anchor the seed to the soil surface thus aiding root penetration. Unanchored seeds, even oriented the right way round, germinate poorly.

Mucilaginous seeds, apart from their virtues listed above, are also more tolerant of humidity changes as they proceed to germination. As Harper (1977) has shown, seeds with copious mucilage such as *Lepidium sativum* and *Camelina sativa* are almost indifferent to the environment's water tension in their germination success. At the other extreme seeds with a smooth testa and no mucilage such as *Brassica napus*, *Pisum sativum* and *Vicia faba* under artificial conditions will scarcely germinate at all with water tensions above 50 cm.

6.4   The Surfaces of Spores

Many fungal spore surfaces are strongly hydrophobic and in some of the cup fungi (members of the Pezizales) spores are violently ejected, still dry, by the impact of rain drops. In the bird's nest fungus genus *Cyathus*, a Basidiomycete, the spores are grouped together in periodioles. When a rain drop strikes into it each peridiole is thrown out of the fungus cup, trailing behind it a mucilaginous thread. Its velocity can carry it up to 2 m; the mucilaginous thread hangs onto a leaf or twig until ripe when the peridiole bursts releasing the spores. Where highly water-repellent spores have been examined under the TEM, such as *Penicillium expansum*, *Neurospora crassa* and *Tricophyton mentagrophytes*, the surfaces are found to be covered with fine rodlets under 1 μm in length. Where these have been examined chemically they are found to consist not of any strikingly hydrophobic substance but 80% protein and the rest glucomannan. Their effective water repellency, however, reinforces the general principle that, provided there is not continuous water contact, the size of the projections is more important than the nature of the exposed surface chemical groups (see Chapter 4).

Other fungal spores may be wettable. The spores of *Phallus impudicus* are covered with a foul-smelling moist mucilage. Flies eat the spore masses which pass through their guts unharmed and the spores are ejected in their excreta over a wide area.

The spores of most mosses are dry and wind disseminated but in the moss *Splachnum*, like *Phallus* above, the spores are mucilaginous and spread by dung flies.

The thalloid liverwort-like plant that crept ashore in the first terrestrial invasion had few problems of reproduction and dispersal. It probably decayed from the centre, and where damaged, fragmented and formed new clones. But as propagating systems became more sophisticated, complicated by the development of meiosis, the respective surfaces, although becoming smaller, had to become more biochemically adventurous. The surfaces of pollen, stigmatic papillae, certain seeds and spores, hyphal tips and the searching haustoria of parasitic plants, with their export and recognition of proteins and other substances are the closest approximation in the plant to the immune response systems of animals. We shall look at these recognition problems again in Chapter 9.

7 Plant Surfaces as a Source of Materials

Both internal and external secretion takes place in plants. Examples of internal secretion are the movement of sugars to the phloem sieve tubes, secretion of oils into the glands embedded in the peel of *Citrus* fruits and the secretion of growth hormones, such as the hormones produced by the shoot apex and other meristems.

External secretion may take place from specialized organs, e.g. foliar glands and trichomes, or over the whole surface of a tissue or organ. We have already seen (Chapter 3) how wax and cutin precursors, synthesized in the epidermis are secreted over the plant surface with the whole epidermis acting as a glandular tissue.

7.1 Roots – Secretion from Root Caps and the Loss of Cells

Roots are covered with a copious discharge of mucilage (Fig. 7–1) from the root cap which, at least in angiosperms, is both a protective and a geoperceptive tissue. The individual cap cells switch from geoperception to secretion and protection as they reach the margin. The amyloplasts lose their starch and, probably stimulated by this, the several hundred dictyosomes in each cell enlarge and throw off numbers of vesicles. These fuse with the plasmalemma, discharging their contents first into the intercellular spaces and subsequently onto the outer surface of the root cap. At the same time the cell wall pectins and hemicelluloses become degraded. The cell contents disorientate, the cells vacuolate and the amyloplasts cease to sediment under their own weight. The secretion is principally a mixture of sugars and sugar oligomers including, at least in some grasses, the rare sugar fucose (a relic of a primitive condition?). This discharge and the partial separation of the cells combine to coat the cap surface in a thick layer of slime in which more or less free but still living cells are embedded. Up to 7000 cells a day may be lost from each maize root tip (Clowes, 1976). As the root is pushed forward by cell elongation, the tip that thrusts against soil particles is not a rigid fragile primary tissue, but a front of free, elastic cells rolling in a thick lubricating jelly (Juniper *et al.*, 1977).

7.2 Secretion of Sugars from the Aerial Parts of Plants

The secretion of sugars from aerial plant surfaces may be localized (e.g. from nectaries) or diffuse. Free sugars occur in leaf washings and leachates. Such loss of photosynthate is often presumed to be expensive in energy terms and may be accidental, although as we shall see, the presence of sugars at plant surfaces may contribute to the maintenance of a healthy flora of saprophytic microbes that help to inhibit pathogens (Blakeman, 1981). The European spindle tree,

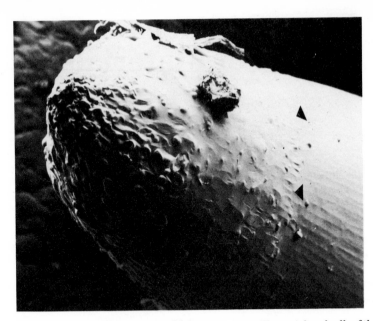

Fig. 7-1   A root cap on a primary root of maize (*Zea mays*). The peripheral cells of the cap have detached, but remain embedded in the thick coat of mucilage. The cap in this species is about 0.5 mm long and its boundary with the elongating zone of the root is indicated by the black triangles. Sputter-cryo. SEM. Courtesy of Dr A. Beckett, Department of Botany, Bristol University.

Euonymus europaea, secretes enough of the sugar dulcitol from its leaf surfaces to form a white crust in dry summers. More often sugars are secreted by specialized glandular trichomes, the nectaries, the function of which is usually to attract insects. Almost all insect-pollinated flowers and many of those visited by bats and birds have floral nectaries at the bases of petals or sepals. The nectar ensures more frequent visits to the flowers and more efficient transfer of pollen to the stigmas.

Extrafloral nectaries also have attractive functions. In the sub-tropical genera *Cecropia*, *Acacia* and *Ipomea*, and probably other genera also, nectaries are associated with specialized structures, such as the swollen hollow bases of thorns in bull's horn *Acacia*, which are inhabited by ants (see Chapter 8).

Carnivorous plants also secrete sugars from glandular trichomes, as we have seen in Chapter 5, as part of the attraction and trapping mechanism and the same glands may also secrete water and enzymes for digestion of the prey and absorption of the degraded proteins.

Nectar-secreting cells are of various types and the gland may have both subepidermal and epidermal components. The sugary secretion is discharged directly onto the surface or into canals or crypts below the surface, as in the various types of extrafloral nectary in the genus *Ipomea*. Typically the nectar-secreting trichomes are multicellular filaments surmounted by a globular apical cell from which the nectar is secreted. At first the nectar collects between the cell and the cuticle. Then it exudes through pores or ruptures in the cuticle when sufficient pressure builds up beneath.

Nectar is predominantly a mixture of the sugars sucrose, fructose and glucose, with small amounts of amino acids and aromatic compounds which may attract certain insects and account for the differences in the flavours of honeys. The sugars are supplied to the gland from the nearest sieve element, probably through symplastic connections, the plasmodesmata (Carr, 1976).

7.3 Secretion of Enzymes

The glands of most carnivorous plants are now known to secrete proteolytic enzymes which degrade entrapped prey. Only in *Darlingtonia* is there no evidence yet of digestive enzyme secretion from the plant itself; digestion of the insect protein is here thought to be performed solely by micro-organisms resident in the pitcher, or brought in on the prey.

A typical carnivorous gland, that of *Dionaea* is depicted in Figure 5-4. This gland is capable of secreting both enzymes and water, as separate processes, and of absorbing the partially degraded products of the entrapped insects. In *Dionaea* a wide range of nitrogen-containing molecules is capable of triggering the active secretion process, and Darwin (1875), who discovered this induction in his experiments on Venus' fly-trap, used small pieces of steak. The most effective of these secretogogues, it has now been discovered, is uric acid, which is copiously produced by flies and some other insects when trapped (Robins, 1976). Such responses to stimuli are not dissimilar to many animal secreting systems.

7.4 Secretion of Salt

Species of plants in several genera, of which the best known are *Limonium* and *Tamarix*, have specialized glands on their aerial surfaces which secrete sodium chloride (Fig. 7-2). In *Atriplex* simple trichomes on the leaf surfaces accumulate salt, then die or break down allowing the salt to be blown or washed away from the surface. These mechanisms may enable them to colonize soils of high salinity. In salt-laden soils such a crust can develop on the living leaf surfaces that both they and the fallen litter are almost fireproof, and for this reason *Tamarix* is sometimes grown as a firebreak between rows of more flammable trees. In the totally submerged marine angiosperms *Zostera* and *Thalassia* the salt eliminating process takes place over the whole epidermal surface.

Specialized salt glands usually comprise two groups of cells (Fig. 7-2) the inner, vacuolate collecting cells (V) and the outer secretory cells (O). The outer cells, of which there may be six or more, bear a striking resemblance to the enzyme-secretory cells of a carnivorous plant. A perforated cuticle (C) covers the outer surface of the salt gland. As in the enzyme-secreting glands of *Drosera* and many other glands, it also extends downwards between the secretory cells and the epidermal cells, separating the secretory cells from the collecting cells; one of the few instances where a true cuticle is developed within a plant tissue. The internal cuticle is perforated by plasmodesmata, via which salt accumulates in the collecting cells and is pumped to the secreting cells. From the secretory cells the salt emerges through the cuticular pores.

As seen in section 5.2.3 the glands on the surfaces of cotton may help to degrade certain viruses. These glands are very like those of *Avicennia* (Fig. 7-2) which secrete NaCl and the chalk-secreting glands of *Plumbago capensis*, but the cotton glands discharge a mixture of Mg^{2+}, K^+ and Ca^{2+} ions probably as the carbonates or bicarbonates and resulting in a pH on the surface of between 7.4 and 10.7.

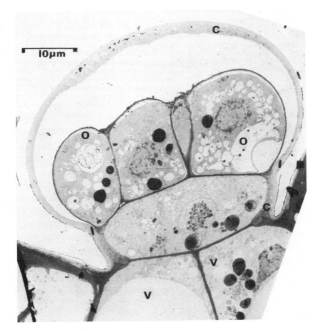

Fig. 7-2   A section through a salt gland of the mangrove *Avicennia resinifera*. (C) the cuticle, (O) outer secretory cells, (V) inner vacuolate cells. TEM. Courtesy of Dr W. W. Thomson, Department of Botany and Plant Sciences, Riverside, California.

7.5   Secretion of Metals

If radio-labelled heavy metals, such as zinc or lead, are added to the soil in which plants as diverse as peas (*Pisum sativum*) and *Pinus sylvestris* are growing, they soon appear as small particles thrown off the leaf surfaces. Particles below 1 μm in size seem to detach spontaneously, larger particles seem to be whipped off by the agitation of leaves in the wind. It seems that a substantial fraction of the trace metal load in the atmosphere may be due to plants growing in metal-rich soils.

7.6   Secretion of Water

Guttation, or the discharge of liquid water from plants, is less frequent than transpiration, but occurs in many species when the plant is saturated with water and the air with water vapour. After a warm, damp night drops of water can be seen hanging on the tips of leaves. Each drop slowly increases in size until it falls off and is replaced. The drops are discharged from specific areas, usually from the extreme tips of most grass leaves, from every leaf tooth in *Alchemilla* and from the ends of the seven main veins in *Tropaeolum majus*. Each drop is secreted from a special gland, the hydathode. Sometimes hydathodes are little more than a cleft in the epidermis. More often a pore is formed from a modified stomatal complex, beneath which a group of non-green, densely cytoplasmic secretory cells, the epithem, differentiate. Water supplied to the epithem from the end of a vein is conducted through it apoplastically, perhaps under pressure. In some species the secretion is active and is stopped by metabolic poisons.

Water-secreting trichomes are found in *Cicer arietinum* and the Brazilian liane *Machaerium*. Specialized unicellular hydathodes may also be developed from epidermal cells and, in *Datura*, water is secreted by apparently undifferentiated epidermal cells. The function of hydathodes has been debated for a hundred years. They may be water regulation devices and it is interesting that they have been reported to *take up* water from dew or rain. However, their chief function is probably to maintain a water flow through the plant when the normal transpiration stream is suppressed and they thereby maintain the uptake of minerals from the soil.

7.7 Secretion of Commercial Waxes

The various waxes in and on the cuticle have been widely exploited by man since antiquity, in candles, soap making, polishes, varnishes, paints and lubricants and in the *lost-wax* method of metal casting (Knaggs, 1947). Industrial society has extended these uses to include electrical insulation, carbon paper and waxed papers, records, pharmaceuticals, cosmetics, dental wax and photographic plates, although in recent years waxes have been replaced by plastics for some of these purposes.

7.7.1 The chemistry of plant waxes

Despite very great technical problems, the major classes of wax compounds were identified in the late nineteenth and early twentieth centuries as long chain hydrocarbons, alkyl esters and free primary alcohols and fatty acids (Fig. 7-3).

Rarer constituents include the flavonoids which give the yellowish powdery coating, the 'farina', to the undersides of the leaves of some Primulaceae, and hydroxy fatty acids in the waxes of the Pinaceae and Cupressaceae. The hydroxy-acids are usually found inter-esterified into ring polymers of about six molecules which are known as 'estolides'; the famed carnauba wax is probably an almost pure polyester of this type.

The wax of a particular species is often constant. But species differ markedly from one another and no one class of compound is always dominant. In *Eucalyptus* species two types of wax occur and β-diketones or free primary alcohols are usually major constituents, either on different species or on different parts of the same species. *Nicotiana* leaf wax on the other hand is mainly composed of alkanes with a high proportion of branched-chain alkanes, while pine wax has secondary alcohols as the dominant class of compounds.

The n-alkanes of higher plant waxes usually have C23–C33 chains, the most common being C29 and C31 and odd carbon-numbered molecules are much commoner than even. The primitive plants of earlier geological periods on the other hand formed even-numbered alkanes. This diversity results from the way in which alkanes are synthesized in any particular plant (see below).

Important commercial vegetable waxes are ouricury wax, from the palm *Syagrus coronata*, waxes from esparto grass and sugar cane, candelilla wax from the Central American *Euphorbia cerifera* (Fig, 7-4), bayberry wax from the fruits of American and African species of *Myrica* and raffia wax from the palm *Raphia pedunculata*.

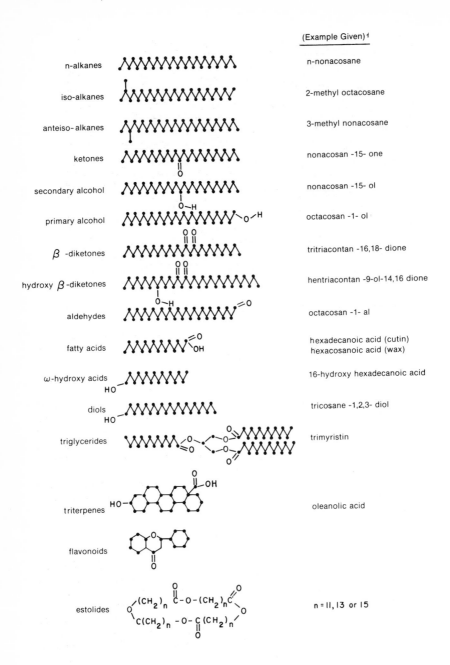

Fig. 7-3   The main classes of compounds found in plant waxes.

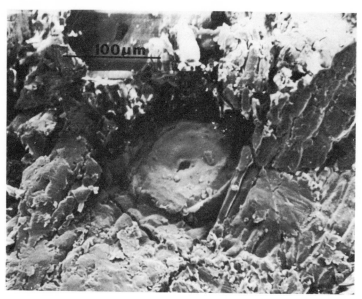

Fig. 7-4   The surface of the stem of *Euphorbia cerifera* (Candelilla) showing the thick wax plates on the surface. The stomatal guard cells are themselves thickly coated and lie in depressions formed in the deep waxy layer. SEM.

The queen of waxes, however, is carnauba wax from the leaves of the palm *Copernica cerifera*, which grows in semi-arid parts of Venezuela and North-East Brazil. Carnauba is a hard, high-melting-point wax which takes a high, durable polish, making it a prime requisite of the best hard car polishes. These properties are due to a high proportion of C55 polymers, polyesters often known as estolides with little alkane or cyclic component. Carnauba is collected by heating leaves with water to melt the wax and two hundred leaves produce each kilogram. This enormous labour makes carnauba an exclusive product, which at about £400 per ton is so expensive, yet so useful, that inferior waxes are added to it as extenders.

7.8   The Chemistry of Cutin

Unlike wax, cutin has no commercial value. Cuticles, isolated from the epidermal cells (Chapter 3; Fig. 7-5) and cleaned of wax consist of the inert polymer 'cutin' which strongly resists acid attack and decay, but which readily dissolves in strong alkalis to produce a mixture of fatty acids and hydroxy-acids, mainly with C16–C18 chains. The most common constituents of cutin are 9,10,18-trihydroxyoctadecanoic acid and 19,16-dihydroxyhexadecanoic acid. The reactive groups of these molecules link and cross link mainly by ester bonds (−CO−O−) (about 75%), but also by peroxide and ether linkages into a high molecular weight three-dimensional network. In practice there is usually an excess of free hydroxyl (OH) groups. Basic dyes also indicate the presence of free carboxyl groups (−COOH).

Fig. 7-5   The cuticular membrane of an *Epiphyllum* sp. isolated by rotting the stem in water; the method used by Brongniart. This view from the inside shows the outlines of the cells of the stomatal complex as a 'cast', like the shed skin of a snake. SEM.

7.9   Resins, Terpenes and Oils

Glandular trichomes are sources of resins, terpenes and volatile oils, some of which have economic importance. The terpene-secreting glands of many solanaceous plants, e.g. the tomato and potato, create hazards for aphids (see Chapter 5) and are responsible for the characteristic and elusive flavour of fresh-picked tomatoes which is rapidly lost when the fruit is handled for marketing. Many aromatic flavourings are derived from glandular trichomes, thymol from thyme, menthol and peppermint oil from *Mentha piperita* and a lemon flavoured oil from *Verbena*. The resinous drug cannabis, which contains the drug tetrahydrocannabinol, is secreted by trichomes on the leaves and flowers of *Cannabis sativa*. The resin ladanum, incorrectly referred to as 'myrrh' in the Bible, is collected from plants of *Cistus ladanifer* by wiping the foliage with a cloth, or by grooming the beards of goats which browse the branches.

In pine needles, in the floral buds or cloves of *Eugenia caryophyllata*, and leaves of *Eucalyptus* aromatic resins or oils are secreted internally into hollow spherical or tubular cavities and ducts, lined with many secretory cells. Similar glands occur in the peels of citrus fruits, but whereas in *Eucalyptus* the cavities are formed by the enlargement of intercellular spaces between the secretory cells (schizogenous glands), those in citrus are formed by disintegration or lysis of a central mass of cells (lysigenous glands).

Other gums and resins originating from internal secretions are exuded onto and collected from plant surfaces. These include the latex-like frankincense from *Boswellia carteri* and the true myrrh, a red gum from *Commiphora myrrha* which has analgesic properties besides its use as an incense. The resins from pines and other conifers which are exuded to the surface as a wound response may be part of a chemical defence mechanism.

Trichomes, as we have seen in Chapter 5, may not only be injurious to insects, but to man as well. The notoriously sticky trichomes of *Nicotiana* (*c.f.* Fig. 5-2) are also very good at capturing inorganic particles, and amongst these may, it is suggested, be radioactive lead, Pb–210. As this decays it produces another isotope, polonium–210, an alpha-emitter. These trichomes, with their various attached particles, tend to persist through the various stages of tobacco manufacture. When a cigarette is lit the sticky trichome-exudate fuses into a plastic-like knot which may glue itself immovably within the body. Alpha-emitters, although having a low energy for penetration and thought therefore to be insignificant outside the body, may be a serious carcinogenic source when glued onto a sensitive internal surface.

7.10   Industrial Uses of Other Surface Products

7.10.1   Fibres from plant trichomes

Non-glandular trichomes rarely achieve any economic importance. Two major exceptions are the long cellulosic hairs formed on the seed coats of cotton (*Gossypium hirsutum*) and *Ceiba pentandra* which yields the fibre kapok. Cotton fibres are unicellular hairs of prodigious length (2 to 6 cm) making them suitable for spinning into long strong threads and hence for weaving fine fabrics. Cotton fabric is known from at least 1000 B.C. and many species of *Gossypium* have been used as sources. Kapok fibre is less strong and is mainly used unwoven in quilts, or as stuffing or padding material in upholstery.

Minor uses of trichomes are known from folklore, such as the practice of lining shoes in winter with the woolly leaves of mullein, *Verbascum lychnitis*. The silicified trichomes of the Brazilian cerrado tree *Curitella americana* make its leaves useful to the Indians as sandpaper. The spines and bristles of several species have found uses as needles, fish hooks and textile finishers (teazles).

7.10.2   Commercial uses of barks

Primitive civilizations have made use of barks as a source of 'paper', for the skin of birch bark canoes and as a medium for sculpture. The Amerindians of Brazil make a non-woven fabric by beating the bark of trees into a fibrous mesh. In modern civilizations the corky bark of *Quercus suber* has been used as a source of tannins for tanning leather and as stoppers for wine and other liquids. Cork when gound up and reprocessed is made into electrical and thermal insulators, flooring material and gaskets. The dried bark from young branches of *Cinnamonum zeylanicum*, (commercial cinnamon) contains about 1% of cinnamon oil. A similar oil containing 55 to 75% of cinnamic aldehyde comes from the bark of cassia, *C. cassia*. The antimalarial drug, quinine and other related alkaloids of medicinal importance are extracted from *Cinchona* bark.

Vast quantities of bark are the by-product of conifer logging operations and after processing, to remove waxes and resins (the bark of Douglas fir contains as much as 10% wax which may have commercial importance), the bark is marketed as a soil-conditioning material used like peat.

7.11   Plant Surfaces and the Origins of Petroleum

A blue atmospheric haze, accompanied by a pleasant 'pine tree' smell of terpenes is a normal characteristic of forested mountainous regions, hence the

Smoky Mountains of Virginia and the Blue Mountains of Australia.

The haze, which has parallels with the man-made smogs of Los Angeles and other urban areas, is caused by photo-chemical oxidation of volatile organic compounds discharged from plants, resulting in an aerosol which scatters blue and transmits red light. A recent suggestion, backed by some experimental evidence, is that epicuticular wax particles become detached during atmospheric electrical discharges, adding to the photo-chemical aerosol. This haze consists of enormous numbers of small, hydrophylic condensation nuclei, exceeding 10^6 ml^{-3} over the Amazon rain forest, down to 10^3 ml^{-3} over the great oceans. Initially as small as 10 nm in diameter these nuclei grow by aggregation up to about 1 μm.

Terpene concentrations vary in rural air from about 2 μg m^{-3} in winter to about 10 μg m^{-3} in summer and about 20 μg m^{-3} in autumn, with a superimposed diurnal cycle peaking at about midday. Went (1974) points out that of 2×10^{11} t of organic material photosynthesized annually, about 0.2% is carotenoid and phytols which probably decompose to terpenoids. Direct terpene production accounts for about 0.5% of total photosynthate, about 10^9 t per annum, giving a total of about 1.4×10^9 t of volatile plant products per annum. Went calculates that there must be 0.3×10^9 t of particulate matter over the world each year, or about 0.25% of a whole year's photosynthates.

Thus, the man-made haze or smog of urban areas is far outweighed in volume by the natural photo-chemical haze derived from hydrocarbon emissions from vegetation. Obviously, not all of this derives directly from the leaf surface, but its diurnal rhythm suggest that a substantial proportion may do so. Some is totally degraded by the sunlight's ionizing properties, but much of it falls or is washed back to earth, to be oxidized or to be protected from oxidation by adsorption onto clay particles. The adsorbed material eventually accumulates in the anaerobic layers of marine and estuarine sediments.

The origin of coals and oils is complex, but Went again speculates that atmospheric aerosols of plant origin may have contributed substantially to petroleum deposits, possibly accounting for some of the aromatic fraction. Inert n-alkanes are major constituents of the cuticular waxes and similar n-alkanes are present in crude oil. In modern plant waxes, and also in marine sediments, the n-alkanes are mostly odd-carbon numbers above 21, whereas those in oil deposits have more equal proportions of odd and even-numbered molecules. C11 and C19 n-alkanes predominate only in the crude oils from the early Palaeozoic formations, possibly derived from the simpler forms of life of that period. In oils of later periods, n-alkanes in the range C20–C30 are found, the longer chain length compounds originating from the waxes of the leaves and fruits of the flowering plants which developed in and dominated the Cainozoic era, rather than from the ferns and their allies from the Palaeozoic.

Under anaerobic conditions whole leaves were occasionally converted into coal, for example, in the paper coals of Russia and Indiana (U.S.A.). Lignite and other low-grade coals were also formed under anaerobic conditions, perhaps rather like the peat bogs of modern times. Modern peats contain about 3% by dry weight of waxes, which can be commercially extracted and about 1% of cutin, together with other organic acids and long chain compounds. These compounds might have contributed to oils, under appropriate circumstances, in the peats of antiquity.

7.12 Leaching of Metabolites from Plant Surfaces

Leaching occurs when water soluble materials are washed off or out of the surfaces of plants by rain, dew or mist. Leaching occurs particularly after dry periods and to a greater extent from very young and senescent leaves rather than from active mature leaves. The materials leached can be diverse and substantial. Apple leaves may lose 25–30 kg of potassium, 9 kg of sodium and 10 kg of calcium per hectare per year, as well as sugars, amino acids and phenolics. While some of this material represents a loss of photosynthate, radio-tracer experiments indicate that much of this leachate is not lost, but recycled via the roots.

Leachates have an important role in influencing the numbers and species composition of plant surface micro-organisms. Sugars and amino-acids stimulate spore germination and are used by them for growth, but the phenolics are toxic to micro-organisms and specific inhibitors of spore germination occur in leaf washings. Plant surfaces can exert controls on the phylloplane flora (see Chapter 8) in indirect ways. Nutrients in leachates encourage a healthy population of benign, saprophytic microbes, now widely recognized as important in resisting pathogens. The natural flora is antagonistic to pathogens, but when crops are treated with fungicides such as benomyl this defence is lost. Modern approaches to pathogen control are just as likely to include the spraying of antagonist organisms or nutrients which encourage their growth to plant surfaces as the application of fungicides, which do not discriminate between the beneficial and the pathogenic.

Leachates frequently have a phytotoxic or allelopathic role, preventing the colonization of herbs or shrubs beneath a canopy. Muller (1966) showed that drips of fog intercepted by *Eucalyptus globulus* leaves contain low concentrations of chlorogenic, *p*-coumarylquinic and gentistic acids which suppress germination and growth of plants beneath the trees. Leachates from *Adenostoma fasciculatum* of the Californian chaparral have several phytotoxic phenolic compounds. *Quercus falcata* leachates contain salicylic acid and *Erica scoparia* of the Iberian peninsula can discharge ten water soluble phenolics, including several common to the leachates from *E. globulus*. Many of these compounds are known also to be active against micro-organisms.

In addition, Muller has shown that allelopathy can be due to volatile emissions from plant surfaces, the best known instances being *Salvia leucophylla* and *Artemisia californica*. Cineole and camphor have been identified from air samples taken in the vicinity of *Salvia* bushes. Both these compounds have strong toxicity to seed germination and effectively suppress competition from other plants within 1–3 m. Second-hand allelopathy is reported from parts of Australia. Around Canberra, chrysomeliad and scarab insects feed on the leaves of *E. globulus* ssp. *bicostata*. Their frass apparently inhibits the germination and growth of seedlings under the canopy, not only because of the concentrations of inhibitory substances already present, but also because new suppressant compounds are synthesized.

7.13 Plant Surfaces as a Source of Nitrogen Compounds

Nitrogen-fixing algae and bacteria grow widely in the phylloplane, both as free-living organisms and in lichens. In British conifer forests 5–15 kg N ha^{-1} may be fixed annually chiefly during the spring and autumn. Although this is an order of magnitude lower than the rate of fixation by alders (*Alnus* spp.) or legumes, the input of nitrogen may be significant in the nutrition of non-fixer species such as conifers in poor soils. A proportion of the nitrogen is recycled by successive generations of phylloplane micro-organisms, but much of it eventually reaches the soil in leachates, leaf litter and the faeces of herbivores, where it becomes available for root uptake. Some phylloplane nitrogen-fixers even continue to fix nitrogen in the rumen of herbivores.

Root nodules in the Leguminosae, and also in *Alnus*, are well known for their symbiotic relationships with nitrogen-fixing bacteria (Fig. 8-4), which accounts for their success as pioneers on nutrient deficient soils. Less well known is the symbiosis of bacteria and plants in the leaf nodules of tropical and sub-tropical species of the *Myrsinaceae* and *Rubiaceae*. Loss of the bacteria has serious consequences for these species resulting in dwarfing, chlorosis, leaf deformation and suppression of flowering. Nitrogen-fixing may not be the most important function of these nodules, however, since nitrogen applications do not always relieve the symptoms.

7.14   Survival of Cutin

Cutin is a profoundly inert substance, capable of withstanding immersion in strong acids (see Chapter 2). Cuticle fragments easily survive digestion in the gut of herbivores and are, therefore, often used as a means of identifying their food plants. Cutin rarely accumulates in well-aerated or alkaline soils where it can be hydrolysed by micro-organisms, but it persists for millennia in the acidic anaerobic environment of peats. More striking still is its survival in plant fossils where the cutin remains essentially unaltered for millions of years, forming a permanent record of the epidermal structure long after all trace of cell walls has been lost (Fig. 7-5). Fossil plants with cuticles were probably land plants. Lack of a cuticle could, however, suggest algae, mosses or liverworts. Alga-like plants possessing a cuticle, such as *Zosterophyllum* from the Australian Silurian, probably represent a transition between lower and higher plants. *Rhynia*, from the Devonian cherts of Aberdeenshire, has both a cuticle and recognizable stomata, which must of necessity have developed parallel with the cuticle. *Spongiophyton* superficially resembles algae or liverworts, but has an 80 μm thick cuticle with air pores 200–300 μm in diameter recalling those of the present-day liverwort *Marchantia* (Fig. 3-1). Many ancient plants are known almost solely from their fossilized cuticles, which may be released from the rock matrix by treatment in hydrofluoric acid.

In modern industrial processes this inertness of cutin can sometimes cause difficulties. Tomato skins from canneries, for example, can accumulate in sewage treatment plants, eventually blocking the filtration systems. They may need expensive alkaline hydrolysis before disposal.

8 The Plant Surface as a Habitat

8.1 The Leaf Surface Flora (Green Plants)

Epiphytes are plants which grow, often high in the forest canopy (but see Fig. 8-1), attached to and entirely supported by other plants. Epiphytes occur in all moist forest, from the tropics to the tundra, but achieve their greatest diversity in the tropical rain forest.

Epiphytes trade the advantage of higher than ground light intensities for an environment where soil is scarce or absent and the supply of water and nutrients is erratic. Epiphytes must, therefore, tolerate complete or partial dehydration, or avoid periods of water shortage by storage or conservation of water and have distinctive methods of absorbing it (Fig. 4-11). Many of the vascular epiphytes such as epiphytic Bromeliaceae and Cactaceae are apparently derived from semi-arid or desert ancestors.

The Pteridophytes and angiosperms are abundant in the epiphytic flora of the tropical forests, but not in the temperate forest. Mosses, hepatics and lichens are ubiquitous. The probable explanation for this is that, in the moist tropics, epiphytes enjoy an almost uninterrupted growing season, whereas those in temperate regions face, during winter, longer periods of physiological drought and, during summer, much greater variation in humidity and water supply. While many non-vascular plants can tolerate periods of substantial or total dehydration, vascular species, with few exceptions, are unable to do so.

Epiphytes, aided by seeds or spores like dust, often root directly into suitable apertures in the surfaces of their hosts or into soft bark or coverings of accumulated organic debris. However, many augment their soil supplies by cup or bracket-shaped arrangements of leaves in which organic litter and soil can collect. Such soil baskets are supplied with a network of roots and one of the most spectacular of these is *Dischidia* (see p. 76). Many species of ants are associated with the root systems of epiphytes and they probably play an important role in their nutrition by collecting particles of soil. However, the leaching of this soil by rainfall must be almost continuous and it may have more value as a water reservoir than as a source of nutrients.

The *Tillandsias* (bromeliads) (section 4.3 and Fig. 4-11) and many species of orchid live on branches in the uppermost reaches of the canopy of forests where supplies of nutrients must be very meagre. Some are capable of surviving on telephone wires or even in open locations where there would be no possibility of obtaining nutrients from leachates. Plants in such locations may obtain nutrients from ants and from the occasional invertebrates and micro-organisms which die on their surfaces, and from their excreta.

Plants which live primarily on *leaf* surfaces, the epiphyllae, are characteristic of tropical, sub-tropical and montane rainforests. The epiphyllae depend on establishing themselves on long-lived evergreen leaves and on an environment with a constant high humidity, regular rainfall and mist and a high ambient temperature. The epiphyllous flora of rainforests is chiefly composed of algae such as *Trentepohlia* and *Phycopeltis*, lichens and leafy liverworts and a few specialist mosses such as the general *Crossomitrium* which occur chiefly in South America, and *Taxithelium* of Malaysia. Interestingly, whereas the epiphyllous liverworts and mosses are rarely parasitic, most of the algae are semi-parasites, *Cephaeuros virescens*, the algal component of the epiphyllous lichen *Strigula*, causes a virulent disease of tea (*Camellia sinensis*) and other plants. The lower surfaces of floating angiosperm leaves may also be host to a range of epiphyllae (Fig. 8-1).

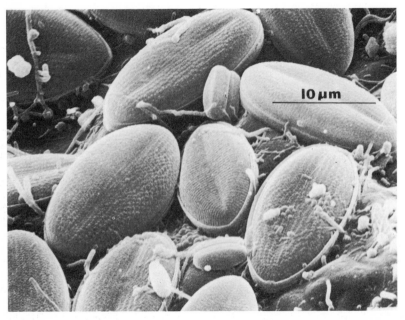

Fig. 8-1   The underside (abaxial) leaf surface of the aquatic angiosperm *Hydrocotyle verticillata* almost completely covered with diatoms (*Cocconeis* sp.), microorganisms and aquatic fungi. SEM. Courtesy of Dr N. D. Hallam, Department of Botany, Monash University, Australia.

Since leaf surfaces are generally smooth and offer little opportunity for rooting or the accumulation of soil, vascular plants can rarely germinate and establish themselves there, although they may be able to do so briefly when a dense community of epiphyllae develops. For similar reasons, mosses which colonize stems and twigs are able to spread to the leaf surface, but cannot usually establish themselves there.

The leaves of tropical rain forest trees often have a pronounced narrow point or 'drip tip', which is a feature rarely found in temperate regions or in the drier regions of the tropics. For long these have been regarded as an adaptation to increase the rate of drainage of the leaf surface, possibly keeping the leaf surface free of epiphyllae, or suppressing the germination of spores and increasing the vigour with which spores and insect eggs are washed off. However, there is no evidence that plants without drip tips, or from which the drip tips have been removed, are less prone to colonization by epiphyllae and no completely convincing explanation for them has yet been suggested.

8.2   The Leaf Surface Flora (Fungi and Bacteria)

Plant surfaces have long been known to play host to fungi and bacteria such as the saprophytic yeasts which ferment wines and give them flavour, and pathogenic rusts and mildews such as *Puccinia* and *Phytophthora*.

Up to 10% of the leaf area may be occupied by fungi, but rarely more than 2% except under very favourable conditions. Conifers and other evergreens may accumulate populations so large that they materially alter the properties of the leaf surface and interfere with physiological functions. Layers of micro-organisms up to 22 μm thick may be found on tropical plants.

Only quite recently, however, has the leaf surface or phylloplane achieved recognition as a well-defined habitat, occupied by specialized communities of non-pathogenic microbes. Population size depends strongly on the availability of moisture and nutrients. Sources of nutrients include leachates, secretions and exudates, soil particles, pollen, dead and inactive spores, aphid honeydew and ions and solutes in rain-water. Plant surfaces are remarkably leaky. Dew and canopy-throughfall contain amino acids, carbohydrates, phenolics, organic acids, antibiotics and growth regulating compounds and inorganic ions. Young and senescent leaves leak most, perhaps reflecting the larger quantities of solutes present in their tissues.

The fungal flora is wide-ranging. In the continuously humid conditions of the moist tropics fungi such as the sooty moulds can make uninterrupted vegetative growth. Elsewhere the phyllosphere is a harsh, fluctuating environment with periods unfavourable for growth. *Sporobolomyces*, *Aureobasidium* and *Cladosporium* have resistant vegetative structures, or may revert to a yeast-like habit when conditions are sub-optimal for growth. Invader organisms such as *Epicoccum* and *Stemphylium* are unable to grow successfully on healthy leaves, but their numbers increase rapidly when leaves are damaged or become senescent. The potentially pathogenic *Botrytis* and *Alternaria* can exist as saprophytic epiphytes before turning parasitic, but the powdery mildews (*Erisyphe* spp.) are more specialist and make little saprophytic growth. The rusts (*Puccinia* spp.) are also obligate parasites, but many have specialized further to the point where they can only invade host plants of a single species. The ascomycete *Vizella* species may complete their whole life cycle in the thick cuticles of genera such as *Banksia* and *Olearia*.

In *Discaria articulata*, a South American genus, even the outer stomatal chambers are so colonized by fungal hyphae as to be virtually sealed. These colonies do not seem to be pathogenic since no cellular penetration was observed and, although they must reduce transpiration rates, they must also hinder gas exchange. The coprophilous species, *Pilobus*, *Sordaria* and *Coprobia* do not live on a plant surface, but may need to lodge there before germination of the spores in faeces.

Under sycamore (*Acer pseudoplatanus*) trees which are growing downwind from coking plants and brickworks the soil contains high concentrations of sulphur. These high concentrations in the soil appear to derive, at least in part,

Fig. 8-2   The adaxial surface of a sycamore (*Acer pseudoplatanus*) heavily contaminated with atmospheric pollution deposits (APD), largely soot, from a nearby coking plant in S. Yorkshire. The APD's are colonized by fungal hyphae and in some cases by yeast-like bodies. SEM. Courtesy of Dr M. Wainwright, Department of Entomology, Sheffield University.

from the phylloplane cycling by fungi and bacteria, of industrial pollutants. Killham and Wainwright (1981) have shown that, in their study, each leaf held about 50 mg of pollutant. There was sufficient carbon, mostly in the form of soot, and nitrogen to support the growth of fungi, and reduced sulphur for bacteria. The phylloplane organisms probably include the fungus *Fusarium solani* and the sulphur-oxidizing bacterium, *Thiobacillus thioparus*. In the SEM micrograph (Fig. 8-2) a combination of micro-organisms can be seen in close contact with the particulate pollutant. From these leaves, probably mainly by rain wash, a shower of the now soluble relatively non-toxic sulphate ion descends onto the soil and may further be modified by the action of soil micro-organisms. Thus a combination of phylloplane micro-organisms solubilize, modify and help to disperse what is potentially a dangerous pollutant and convert it, in part, into an essential nutrient for protein synthesis.

Bacteria of many species inhabit the phyllosphere and because of the rapidity with which they can grow and multiply they are often the primary invaders. Many phylloplane bacteria contain carotenoid pigments which have been interpreted as a defence against the lethal effects of high light intensities. Like fungal spores, bacteria are distributed by wind and rain-splash. Few bacteria develop resistant spores, however, though their cells often survive long periods in adversity. Bacteria chiefly penetrate via wounds or natural openings such as stomata and nectaries. For this reason wind-driven rain is an important factor in their epidemiology, carrying the cells into these openings when the surface becomes flooded with water.

Phylloplane microbes are capable of using wax and cutin as carbon sources and often leave tracks or imprints where wax has been dissolved in their vicinity, sometimes revealing details of cuticle structure beneath. The enzyme systems involved are not understood, though they appear to be secreted by the advancing hyphal tip in filamentous fungi. Cutinolytic activity is equally poorly understood. Although the smooth-edged penetration holes produced by pathogens suggest an enzymic lysis of cutin, it has generally been assumed that the mechanism is purely mechanical. Nevertheless, the cuticle is an effective barrier to penetration, especially from bacteria which, as we have seen, chiefly invade through natural apertures. *Erwinia amylovora*, the bacterium responsible for 'fire blight' of apples and pears, infects via the nectaries, stigma surfaces and other uncutinized areas of flowers and is vectored by bees. *Pseudoperonospora* and *Plasmopora* invade leaves by means of motile zoospores capable of locating and encysting above the stomatal apertures, subsequently penetrating the open or closed stomata via a penetrating hyphal thread. Besides the obvious consequences of infection, such zoospores may interfere with gas exchange.

The manipulation of populations and species composition of the phyllosphere flora offers the potential to control the incidence of crop diseases without the undesirable side effects of resistance and environmental pollution which are associated with application of pesticides to crops. The technique of biological control of microbial pathogens is in its infancy and major practical difficulties must be overcome. The blanket application of fungicides to crops can have the undesirable consequence of eliminating the normal saprophytic flora, thereby providing the pathogen with a clear pitch. Regular supplies of nutrients to the phylloplane may be sufficient to ensure a healthy population of antagonist saprophytes and a moderate leaching loss may be an essential component of a plant's defence strategy.

8.3   Nitrogen Fixation on and in the Leaf Surface

Biological nitrogen fixation is a property specific to a few bacteria and blue-green algae and some of those puzzling prokaryotes, the actinomycetes. Blue-green algae capable of fixing nitrogen are found on, and occasionally in, a wide range of plant surfaces (Ruinen, 1975) and in symbiotic associations. The best known include *Nostoc* species on and sometimes in the cells of the excretory glands and certain other cells at the base of *Gunnera* leaves. Others include *Anabaena* on the free-floating fronds of the fern *Azolla*, *Richelia* endophytic in the diatom *Rhizosolenia*, other blue-greens on and in the coralloid roots of cycads such as *Encephalartos* and *Macrozamia*; *Hapalosiphon* on the moss *Sphagnum*; *Nostoc* species on the liverworts *Blasia* and *Cariculania*; *Nostoc* species again in several lichens such as *Peltigera rufescens* and *Collema tenuiforme* and *Calothrix* with the thallus of the green alga *Enteromorpha* (Nutman, 1976).

It is now widely believed that the conventional blue-green algae entered into symbiotic associations with eukaryotic cells to evolve eventually into the chloroplast. The first indications of nitrogen-fixing blue-greens penetrating *into* living cells of eukaryotes have now been found (Silvester, 1976) and it is tempting to speculate what the future holds for the first fully integrated endosymbiotic nitrogen-fixing organelle!

In the fern *Azolla filiculoides* the apical meristems are infected with *Anabaena axollae* and the blue-green grows in unison with the fern tip. As leaf primordia are formed, filaments of the alga become trapped within the developing cavities in the dorsal lobes of each leaf. The alga lives only in this micro-environment and dies when the leaf senesces. It maintains its continuity by infecting the megasporangium, which, after fertilization and formation of the zygote and still with the infection, develops into the sporophyte (Ashton and Walmsley, 1976).

In the *Gunnera–Nostoc* association, so effective is the nitrogen fixation that all the nitrogen needs of this angiosperm can be supplied. *Gunnera* is a significant pioneer species in New Zealand and is estimated to fix up to 72 kg N ha^{-1}yr^{-1} (Silvester, 1976).

As seen in section 7.13, blue-green algae on the surfaces of conifers may make a significant contribution to the nitrogen regime of that forest. But this is only indirect in that probably most of the fixed nitrogen is washed off the plant surface as leachate and is taken up not only by the host roots, but probably also other plants of the forest floor.

8.4   The Fungal Flora of Roots

Many fungi such as *Ophiobolus graminis* (take-all of wheat), and *Fomes annosus* (butt-rot of conifers) live on the root surface of the host, but may only cause actual disease when the environmental conditions are right and the resistance of the host overcome. Mycorrhizal fungi are similarly restricted to specific host plants. They do not usually cause disease symptoms to appear, but exist in long-lived and apparently mutually beneficial relationships with their hosts.

Mycorrhiza are of two kinds; endotrophic mycorrhizas such as those of orchids live mainly within the host and in this study of surface phenomena do not concern us here. Ecotrophic mycorrhizas form more or less complete fungal sheaths around the primary roots and laterals (Fig. 8-3) and penetrate only between the cells of the root cortex. They are found on most forest trees in temperate regions, conifers, oaks (*Quercus*), beeches (*Fagus*), birches (*Betula*), sweet chestnut (*Castanea*), hornbeam (*Carpinus*) and *Eucalyptus*. Many of these trees will grow perfectly adequately in nutrient-rich soils if uninfected, but it now seems that most temperate tree species will do better on nutrient-poor soils if infected with a mycorrhizal fungus.

As can be seen from Fig. 8-3, the fungi form a sheath so completely surrounding the host tissue that it would seem that all nutrients absorbed by the host root must pass through it. The fungal hyphae of beech (Fig. 8-3A) seem only infrequently to break out from the sheath to make connections with the soil mycelium or exploit soil debris, whereas in the larch (*Larix*) there is an interlocked mycelial tangle. These differences are not yet understood.

More than one hundred species of common Basidiomycetes, such as the genera *Boletus*, *Lactarius*, *Russula* and *Tricholoma*, have been identified as forming mycorrhizal associations. Some are widespread, some restricted to one or a few species of tree.

Mycorrhizas of all types, along with the symbiotic bacterial infections described in the next section and symbioses such as lichens are just another of the many stratagems evolved by diverse plants for exploiting and raising the soil status of nutrient deficient soils.

8.5   The Nigrogen-Fixing Bacterial Flora of Roots

True bacteria, also with the ability to fix nitrogen, but free-living on soil surfaces or leaf litter are immensely important, particularly in forest soils where little light reaches the ground. But they are not strictly part of a plant surface's flora.

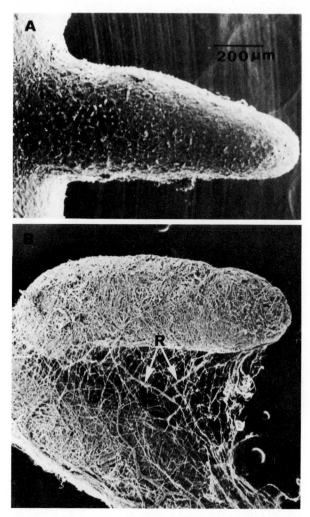

Fig. 8-3 Mycorrhiza on tree roots. **A** On a young root of beech (*Fagus sylvatica*) the fungal sheath is more or less complete around the root, but few mycelial connections can be seen extending from the sheath to the soil. **B** The basidiomycete *Suillus grevillei* on a young root of Japanese larch (*Larix leptolepis*) which has formed a complete sheath with extensive mycelial connections, some of which are beginning to form rhizomorphs (R). SEM both. Courtesy of Dr Jane Duddridge, Department of Agricultural Sciences, Oxford University.

In *Alnus glutinosa* (the European alder) the micro-symbiont *Frankia alni* which is an actinomycete-like organism forms hyphal clusters within the host cells. Similar filamentous prokaryotes form similar intercellular nitrogen-fixing associations in the roots of many species of *Ceanothus* in the western United States.

However, the legume-*Rhizobium* nodule is the most studied of all the nitrogen-fixing symbioses (Fig. 8-4), although it is no longer true to say that rhizobia are confined to Leguminosae or required to live on the host's root surface.

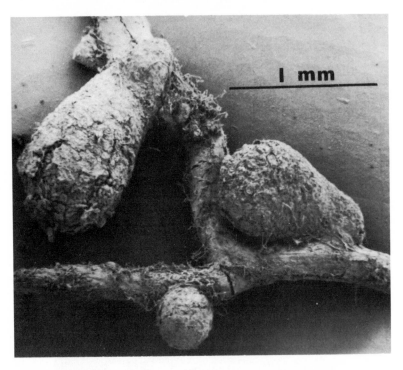

Fig. 8-4   Nodules of the nitrogen-fixing *Rhizobium* on the roots of the clover *Trifolium repens*. SEM. Courtesy of Mr D. Kerr, Oxford.

How does the *Rhizobium* bacterium know which root to infect? It has been suggested that specific lectins such as phytohaemagglutinin might be responsible for binding rhizobia to the root hair. Apparently lectins from certain hosts react only with those rhizobia able to nodulate from that host. However, not all rhizobia capable of nodulating that host will react. The root hairs of the legumes and their nodulating rhizobia may have common surface antigen determinants and the multivalent lectins may act as a bridge between the bacteria and the host; it is possible that cellulose microfibril formation by the rhizobia may also be involved. An alternative suggestion is that the cell wall lipopolysaccharides, rather than the exopolysaccharides are the bacterial component which reacts with the lectin. However, recognition, as we have seen with pollen on stigmatic surfaces, is only the first step to root hair invasion. Infection thread growth and nodule initiation must then follow. Although a great deal of work has been directed at these stages in the last few years they are still imperfectly understood.

As we have seen, nitrogen fixation is not limited to root surfaces nor to bacteria or to the Leguminosae, but it is still probably true to say that of the plant surface nitrogen-fixers, *Rhizobium* species on the roots of Leguminosae are the most significant in northern agriculture. However, *Spirillum lipoferum* infects and fixes nitrogen in the roots of some Gramineae including maize and it is probably in the fodder and forage grasses and the cereals that the next major advances in natural nitrogen-fixation will occur.

8.6 The Non-Nitrogen-Fixing Bacterial Flora of Roots

Apart from the nitrogen-fixing bacteria, another group of bacteria of the *Pseudomonas fluorescens-putida* group, which rapidly colonize the surfaces of the roots of potato, sugar beet and radish, may also cause increases in crop yields. Field tests have shown that artificially inoculated crops may increase their yields by up to 144%. These plant growth-promoting rhizobacteria (PGPR) may increase yield by, as we have seen with some fungi (Chapter 5), interfering with potentially deleterious rhizoplane fungi and bacteria. It is now thought that they may do this by depriving the native and possibly hostile microflora of iron. PGPRs apparently produce extracellular siderophores (microbial iron transport agents) which complex and thus render unavailable the environmental iron (Kloepper *et al.*, 1980).

8.7 Animal Guests

The word 'domatia' (little houses) here means various structures on plant surfaces adapted for guests, both animal and plant, which are of service to the host. These homes may be depressions, pockets, micro-marsupia or tufts of hairs, commonly in the principle vein axils, but occasionally in almost any vegetative part of the plant. They may even take the form of swollen hypocotyls, stems or stipular thorns perforated with holes. Domatia are most common on woody plants of humid tropical regions, rarer in colder regions and, so far, apparently absent from dry deserts.

Domatia are now interpreted as normal morphological developments of use to the plant as dwellings for commensals. The ant plants of Malaysia and Papua, mainly of the Rubiaceae have swollen hypocotyls and stems which, although they may be secondarily modified by species of ants, are formed on uninfected plants. The extensively swollen, tuberous hypocotyls of *Myrmecodia* and *Hydnophytum* are penetrated by a labyrinth of chambers and passages inhabited by many genera of ants.

Ant-epiphyte associations are widespread in the plant kingdom and there may be up to 200 species, concentrated in the Rubiaceae, but also in the Asclepiadaceae, Bromeliaceae, Orchidaceae and the fern family Polypodiaceae.

There is no doubt now that in, for example, *Myrmecodia* (Rubiaceae) and *Tillandsia* (Bromeliaceae), the presence of the ants in the surface chambers aids the nutrition of the plant. The ants' faeces, food materials and decaying bodies will contribute to the nitrogen/phosphorus/potassium supply of the otherwise ion-limited epiphyte. There is now good evidence that ^{32}P can be taken up from these chambers and, cations and anions apart, there are suggestions that both the water economy and the CO_2 balance of the plant may be improved by the ants' presence. What is more, in some way that is totally unclear, the presence of the ants of the genus *Pheidole*, inhabiting the curved petioles of *Piper cenocladum*, induces the plant to produce food bodies. Virtually no food bodies are produced in the absence of ants.

One of the most spectacular of these ant-epiphytes is *Dischidia* (Asclepiadaceae) (Fig. 8-5) with its flower pot. These pots are commonly inhabited by adventitious roots from the parent plant, ants, ant-brood, fungi, micro-organisms and a range of ant-introduced and casual debris. It is recorded that adventitious root penetration is stimulated by the presence of detritus in the pitcher. The parallel development of such pitchers which do not trap insects is interesting to compare with those different genera of pitcher plants which do both trap and kill insects for their own use (Chapter 5). So in *Dischidia* we have a plant which is an epiphyte in its own right, but in turn provides at least a temporary home for members of the animal kingdom.

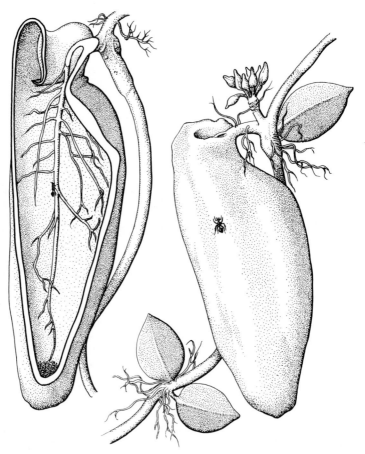

Fig. 8-5   The pitcher of *Dischidia rafflesiana*, a S.E. Asian climbing epiphyte. Drawing by Mrs Rosemary Wise, Botany School, Oxford University.

9 The Wound Response, Grafting and Chimeras

9.1 The Wound Response

When plants are wounded internal tissues become exposed and begin a new function as the plant surface. The behaviour of plant tissues after wounding is so characteristic that it is often known as the 'wound response'. Wounding strongly stimulates mitotic activity, and even differentiated cells may dedifferentiate before dividing one or more times. However the guard cells do not divide after wounding, indicating that their differentiation is not reversible, and pith cells do not normally respond either (Fig. 9-1). Cells close to a cut surface show increased respiration, and may rise slightly in temperature (*ca.* 0.1°C).

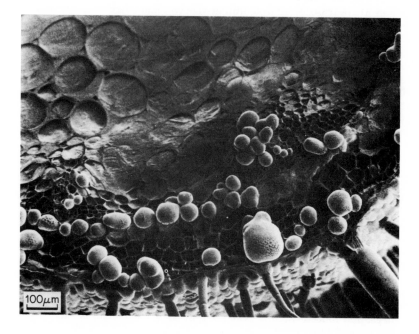

100μm

Fig. 9-1 The transversely cut surface of a tomato stem (graft scion) 2 days after excision, showing wound callus cells, the products of cell divisions in the vascular tissues and outer cortex. At this stage few new cells have originated in the inner cortex, and the pith appears to lack meristematic potential. SEM. Cyrogenic specimen stage.

Ultrastructural changes occur, such as proliferation of the endoplasmic reticulum, enlargement of nuclear pores and reduction in the numbers of dictyosomes, and similar changes also occur in cells subjected to chemical injury. The wound response is transmitted to cells some distance from the point of injury, and was once thought to be due to the action of specific wound hormones, for which traumatic acid, 2 dodecenedioic acid, $HOOC(CH_2)_8CH=CHOOH$, liberated by the lipoxidase mediated oxidation of unsaturated fatty acids, was once regarded as a likely candidate. Traumatic acid can stimulate cell division, and does induce tumorous growths in tomatoes, but other plants are apparently immune to its effects, despite exhibiting a normal wound response, and it is now generally thought that there is little evidence for a specific effect (Lipetz, 1970). Lipetz speculated that lysosomes may be involved, liberating from damaged cells materials which are normally sequestered, and which act as growth hormones. The effect of the wound response is to produce a wound periderm, perhaps of 6 to 10 cell rows close to the damaged surface, with the new cell plates aligned periclinally. Cells exposed at the wound surfaces are prone to desiccation, since they have lost the protection of antitranspirant extracellular coatings of cutin and suberin, and they are also vulnerable to pathogen attack. Suberin and then lignin, possibly derived from a common precursor, are deposited in the outer cell walls of the wound periderm. The lignin produced is identical to that in normal cells. Wax metabolic pathways are also increased leading to the secretion of waterproofing layers of cutin and wax onto the wound surface. The free fatty acids used as raw materials for this synthesis (see Chapter 3) are transferred from ininjured leaves and other tissues remote from the wound where their synthesis increases, further evidence of some, as yet unidentified, hormonal involvement in the control of the wound response. As the synthesis of lignin and suberin proceeds cell divisions in the periderm cease, and the tissues return to a normal state. However in wounds which are kept moist, protected for example by enclosure in wax or polythene, lignification and suberization of the wound fail to take place, and cell division in the vicinity of the cut surface may continue, producing wound a callus. It appears, therefore, that desiccation of the wound callus cells, or possibly their exposure to oxygen, may act as a counter stimulus to the wound response, bringing about the differentiation of the wound periderm, and the cessation of cell division.

9.2   Grafting of Plant Surfaces

When maintained in a moist environment the new cells which proliferate at a damaged surface, whether the result of a complete severing of a stem, or a grazing wound, are thin walled and highly vacuolate. Seen in surface view in the SEM they appear rounded, and are at first smooth, but may develop surface projections which contain pectin, and possibly also lipids (Fig. 9-2). These cells are receptive to contact with callus cells growing out from another wounded surface, and if brought together the two surfaces will adhere, the pectic projections spreading between them to form a 'middle lamella'. Similar cell–cell bonding by pectinaceous projections can be observed in callus cells in tissue cultures, and indeed there is evidence that the phenomenon is widespread in loosely packed parenchymatous tissues. (Carlquist, 1956; Davies and Lewis, 1981.) This process of initial adhesion is the first step in the formation of a graft union (Yeoman and Brown, 1976; Yeoman, 1982). Grafting is chiefly known as a horticultural technique. Yet it occurs repeatedly in nature, especially in the

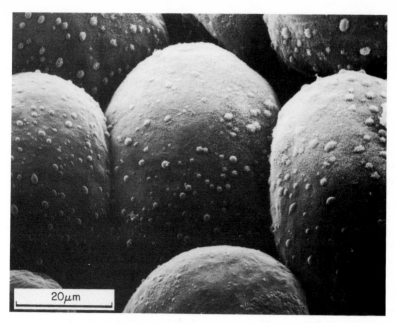

Fig. 9-2 Surfaces of wound callus cells growing out from the cut surface of a tomato stem (graft scion) at 4 days after grafting, showing pectinaceous projections which form a 'middle lamella' when the cells of stock and scion meet across the graft union. SEM. Cryogenic specimen stage.

mechanically and environmentally stable conditions in the soil, between roots of perennial plants which become closely adpressed. This type of grafting, known as approach grafting, can occur between parts of the same individual (autografts), between individuals of the same species (homografts) or between compatible individuals of different species (heterografts). It has the hitherto little understood or investigated significance that assemblages of plants may become physiologically interlinked in a way which offers the potential benefits of a shared supply of assimilates and water, but also has the negative possibility of the systemic transmission of virus diseases between individuals. In a graft between incompatible partners the wound response is extended, and the space between stock and scion fills with a mass of callus, forming a physical barrier between the partners. If the scion has no roots of its own it will eventually die. The wounded surfaces of compatible partners do not continue to proliferate callus cells in an uncontrolled manner. There is evidence that as soon as the cells growing out from stock and scion meet some exchange of information, perhaps a specific recognition event, takes place which has the effect of limiting the callus development, and leads ultimately to the differentiation of the cells into xylem and phloem, forming a functional vascular reconnection. Following this initial recognition event the cell walls of opposing cells in the union become locally eroded, and *de novo* plasmodesmatal connections develop between them. This response is not elicited by contact of the cells with inert surfaces, and is not therefore caused simply by mechanical stimuli, but by a more specific, probably chemical, exchange. The nature of the recognition event or events which take place in grafts, and their location on the cell surfaces, are the subject of intense investigation, but are still obscure. It seems probable that the process can be divided into at least two steps. Intercellular communication at the point of first contact must presumably be due to an exchange of diffusible molecules, capable of traversing thin cell walls since the plasmalemmas are separated by the cell walls at this stage. The development of plasmodesmata, however, because they result in a structural interconnection of the cell membranes across the graft union, may be the location of the definitive exchange of information between stock and scion which leads to acceptance or rejection.

The general topic of cellular recognition processes at plant surfaces is of great importance since it is beginning to explain the highly specific, but previously mysterious interactions between pollen and stigma, between motile gametes in algae, between rhizobia and roots, and between plant and pathogen. Recognition processes may also be relevant to the ways in which cells organize as a highly differentiated community in the plant body. Suffice it to say that although lectins – plant proteins and glycoproteins which have the capacity to bind in a highly specific way to carbohydrate residues on cell surfaces – have been implicated in some recognition events involving plant cells, they do not appear to be involved in recognition processes in the graft (Heslop-Harrison, 1978).

9.3   Chimeras

Although the partners in a graft normally retain their genetic and phenotypic identity, occasional grafts throw up shoots from the graft union which have characteristics intermediate between those of the two components. Such shoots were initially known as 'graft hybrids', but Winkler (1907) called them 'chimeras' after a Greek mythological monster with a lion's head, a goat's body and the tail of a serpent. Chimeras of this type arise from adventitious buds organized out of a mixture of stock and scion cells.

The developing shoots of the form known as 'sectorial chimeras' (see below) may be divided more or less bilaterally into sectors characteristic of each parent, and some variegated varieties of *Chlorophytum* are of this type. Buds developing within the tissue of each parent will be true to the parental type, but buds arising at a junction between sectors will give rise to sectorial chimeras. Occasionally, where a layer of one parent perhaps one or two cells thick overlies tissue of the other, shoots may arise entirely coated with a 'skin' of one parent, but with a core tissue from the other. Figure 9-3 illustrates how this may happen. *+Laburnocytisus adami*, with its epidermis of *Cytisus* and core of *Laburnum* probably arose in this way. Such periclinal chimeras are highly stable; *+L. adami* is over 150 years old, whereas sectorials are unstable. Sectorials have been recorded as arising from grafts beween species of a single genus such as the Bizzarria orange, which is probably a periclinal chimera of *Citrus aurantium* and *C. medica* (Baur, 1930) and between distinct species, as in the chimeras of *Solanum nigrum* and *Lycopersicon esculentum* described by Winkler (1907). Other types, the cytochimeras, can arise as a result of plastid mutations or changes in ploidy in the surface layers of a shoot.

Nineteenth-century botanists noted that the growing point of the plant apex appeared to consist of two (monocotyledons) or three (dicotyledons) germ layers. Although now recognized as an oversimplification (see p. 81), this idea was consolidated by Schmidt (1924) into the tunica corpus theory, in which he proposed that the outer skin of the shoot apex, the tunica, consisted of one

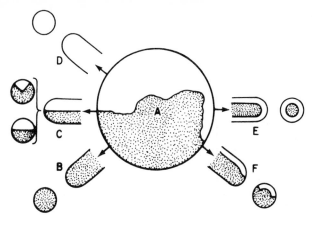

Fig. 9-3 Diagram to show how, from the stem of a sectorial chimera (A), arise pure green (B), sectorial (C), pure white (D), periclinal (E) and mericlinal (F) branches. Redrawn from Neilson-Jones (1969).

(monocot) or two (dicotyledons) periclinal layes of cells in which the cell divisions were only anticlinal, overlying a core, the corpus, in which the cells divided in random orientations. This layered arrangement is clearly seen in Figure 9-4 and accounts for most of the known periclinal chimeras. For example, in both of Baur's green-and-white periclinal chimeras raised from a sectorial chimera *Pelargonium zonale* var. Albomarginatum the tunica consisted of tissue of one parent, and the corpus of tissue from the other. His green-over-white chimera had leaves with a green margin and white centre, but note that this can only occur in the anomalously developing *Pelargonium* 'Freak of Nature', while white-over-green produced leaves with a white margin and a green centre. Winkler's chimeras of black nightshade and tomato showed four distinct forms intermediate between the parental types, and in his analysis the growing points were configured as shown in Fig. 9-5 below.

Winkler's chimeras, although they resulted in considerable permutation of the parental characters, such as changes in leaf shape, fruit and flower size and colour, resulted in no characters which were wholly new. However a more recent periclinal chimera of *Camellia sasanqua* and *C. japonica* resulted in a flower with 6 or 8 styles, in contrast with the parental 3. Clearly, therefore, the tissues of distinct species are not only capable of cooperating to produce a fully functional plant, but in doing so they can produce entirely new characters which might conceivably have adaptive significance.

Cytochimeras having ploidy changes in the tunica or the corpus can often be induced by treatment with colchicine, which prevents the separation of the chromosomes at metaphase, and can result in their fusion into a tetraploid nucleus when the effect wears off. Cytochimeras have helped elucidate the fate of the germ layers in the developing tissues of the shoot. Dermen (1947) showed, using cytochimeras of peach, that the germ layers are not histogens in the sense that each rigidly contributes to a prescribed set of tissues. Dermen's cyto-chimeras, and those of *Datura* studied by Satina *et al.* (1940) clearly demonstrated that the germ layers at the extreme tip of the growing point functioned independently, giving the tunica-corpus theory some validity. It was

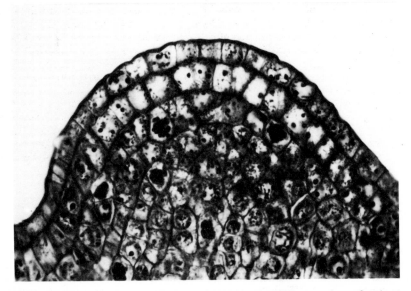

Fig. 9-4 Median longitudinal section of the shoot apex of *Pisum sativum* showing two outer periclinal layers of cells, the tunica, over a randomly orientated corpus.

Fig. 9-5 A diagram of the structures of the growing points of Winkler's periclinal chimeras between black nightshade and tomato, and their effect on leaf shape. Tissues derived from tomato are shaded, and those derived from black nightshade are unshaded.

also established that the outer cell layer, tunica 1, generally contributes only to the epidermis, while tunica 2 gives rise to the cortex but may also form vascular tissues. However the corpus has the capacity to contribute to all of the tissues, although it is usually overlain by tissue derived from tunicas 1 and 2.

Apart from their interest as pointers to the organization of the shoot apex cytochimeras have some horticultural and agricultural usefulness. Certain sectorial and periclinal chimeras, especially those which give rise to variegated foliage, are widely propagated. However periclinal chimeras have the much more interesting potential that a disease-resistant variety of a commercially significant plant, such as the potato, might be achieved by grafting the tunica of a disease-resistant variety (or even of a distinct species) onto the corpus of a susceptible, but high-yielding, variety. Heichel and Anagnostakis (1978) showed that periclinal chimeras of tomato with *Solanum pennellii* could have lower rates of transpiration than either parent, suggesting the possibility of producing drought resistant varieties through chimeras. Although plant breeders are aware of such possibilities it is not yet possible to produce chimeras to order, because although they are known to arise from grafts, they do so with great rarity. An illustration of this is that the list of fully authenticated woody chimeras is still in single figures. However recent work by Mark Holden at Edinburgh has established that it is possible to graft the epidermis, or the epidermis with one or two cell layers attached, back onto the plant from which the tissue was removed, or onto wounded sites on other, compatible plant species. Epidermal grafting can be performed *in vitro* and may offer the basis of a controlled method of chimera production.

9.4 Plant Parasitism

The interactions between the surface of vascular plant parasites and their hosts parallel the interactions between graft stock and scion at least in the sense that the uniting of the two appears to involve recognition phenomena, and results in the establishment of vascular interconnections between the two components. Judged by the criteria usually thought necessary for a compatible graft, namely that the components should be closely-related taxonomically, most parasites would be incompatible with their hosts. Yet parasites are capable, by some unknown means, of overcoming the natural tendency of the host to reject the invasion of its tissues. As might be expected, most parasites are host-specific. The broomrapes (*Orobanche*) as a group are parasites of a wide range of genera, but each species of *Orobanche* is quite specific in its parasitism, *O. purpurea* on *Achillea*, *O. elatior* on *Centaurea scabiosa*, *O. maritima* on *Daucus carota* and so on. Other parasites have more catholic tastes. The lesser dodder, *Cuscuta epithymium* is a parasite of genera as distant as *Erica* and *Ulex*. Clearly the methods used by parasites to avoid or inhibit rejection by their host can in some instances have a fairly general application.

Some plant parasites form cytoplasmic connections with their hosts via plasmodesmata formed *de novo* between them (Carr, 1976). Such connections may be short-lived, as they are in *Cuscuta*, where the host often rapidly blocks them with wall materials. Their significance is not only as pathways of transport for assimilates from host to parasite but also as channels of communication between parasite and host. *Cuscuta* species are known to be able to inhibit flowering and fruiting in their hosts, inducing them to remain in a vegetative state. This inhibition may stem from the absorption of host photosynthate by the parasite, but the possibility exists that the parasite redirects the host's development by means of growth regulators transmitted across the host–parasite interface. Since some species of *Cuscuta* are small enough to lend themselves to *in vitro* cultivation on host tissues they may be an ideal system not only for studies of cell–cell interactions and recognition, but also for studies of growth regulation in plants.

References

Ashton, P. J. and Walmsley, R. D. (1976). The aquatic fern *Azolla* and its *Anabaena* symbiont. *Endeavour* **35**, 39–43.

Baker, E.A. and Holloway, P. J. (1971). Scanning electron microscopy of waxes on plant surfaces. *Micron* **2**, 364–80.

Barber, J. (1977). *Mucilaginous Seeds, Interactions with Micro-organisms.* Supplements to Plant Physiology, *Lancaster* (abstracts) **59**, 35.

Baur, E. (1930). *Einführung in die experimentelle Vererbungslehre.* Berlin.

Bianchi, A. and Marchesi, G. (1960). The surface of the leaf in normal and glossy maize seedlings. *Zeitschrift für Vererblehre* **91**, 214–19.

Blakeman, J. P. (ed.) (1981). *Microbial Ecology of the Phylloplane.* 3rd Int. Symp. Microbiology of Leaf Surfaces, Aberdeen, Sept. 1980. Academic Press, London.

Brodie, P. B. (1842). *Proceedings of the Geological Society* **3**, 592 only. Notice on the occurrence of plants in the plastic clay of the Hampshire coast.

Brongniart, A. (1834). Sur l'épiderme des plantes. *Annales des sciences naturelle Bot.)* 2nd ser. **1**, 65–71.

Carlquist, S. (1956). Wound healing in higher plants. *American Journal of Botany* **43**, 425–9.

Carlquist, S. (1976). Wood anatomy of the Roridulaceae: ecological and phylogenetic implications. *American Journal of Botany* **63**, 1003–8.

Carr, D. J. (1976). Plasmodesmata in growth and development. In: *Intercellular Communication in Plants: Studies on plasmodesmata.* Eds B. E. S. Gunning and A. W. Robards. Springer Verlag, Berlin.

Chibnall, A. C., Williams, E. F., Latner, A. L. and Piper, S. H. (1933). The isolation of n-triacontanol from lucerne wax. *Biochemical Journal* **27**, 1885–8.

Clowes, F. A. L. (1976). Meristems. In *Perspectives in Experimental Biology* Vol. 2, Botanical Symposium of the Society for Experimental Biology (ed. N. Sunderland) p. 25. Pergamon, Oxford and New York.

Cutler, D. F., Alvin, K. L. and Price, C. E. (1982). *The Plant Cuticle.* Linnean Society Symposium Series. Academic Press, London.

Cutter, E. G. (1978) *Plant Anatomy Part 1: Cells and Tissues.* 2nd Edition. Edward Arnold, London.

Darwin, C. (1875). *Insectivorous Plants.* Murray, London.

Davies, W. P. and Lewis, B. G. (1981). Development of pectic projections on the surface of wound callus cells of *Daucus carota* L. *Annals of Botany* **47**, 409–13.

de Bary, A. (1871). Über die Wachsüberzüge der Epidermis. *Botanische Zeitung* **29**, 129–619.

Dermen, H. (1947). Periclinal chimeras and histogenesis in cranberry. *American Journal of Botany* 34, 32–43.

Dickinson, C. H. and Preece, T. F. (Eds) (1976). *Microbiology of Aerial Plant Surfaces.* Academic Press, New York and London.

Dyson, W. G. and Herbin, G. A. (1968). Studies of plant cuticular waxes IV. Leaf wax alkanes as a taxonomic discriminant for cypresses grown in Kenya. *Phytochemistry* 7, 1339–44.

Elleman, C. J. and Entwistle, P. F. (1982). A study of glands on cotton responsible for the high pH and cation concentration of the leaf surface. *Annals of Applied Biology* (in press).

Grimstone, A. V. (1976). *The Electron Microscope in Biology.* 2nd Edition. Studies in Biology no. 9. Edward Arnold, London.

Haberlandt, G. (1914). *Physiological Plant Anatomy.* Macmillan & Co. Ltd., London.

Hallam, N. D. and Juniper, B. E. (1971). The anatomy of the leaf surface. In *Ecology of Leaf Surface Micro-organisms,* p. 3–37. Eds T. F. Preece and C. H. Dickinson. Academic Press, London and New York.

Harper, J. (1977). *Population Biology of Plants.* Academic Press, London, New York.

Heichel, G. H. and Anagnostakis, S. L. (1978). Stomatal response to light of *Solanum pennellii, Lycopersicon esculentum,* and a graft-induced chimera. *Plant Physiology, Lancaster* 62, 387–90.

Heslop-Harrison, J. (1978). *Cellular Recognition Systems in Plants.* Studies in Biology no. 100. Edward Arnold, London.

Holloway, P. J. (1968). The effects of superficial wax on leaf wettability. *Annals of Applied Biology* 63, 145–53.

Holloway, P. J. and Baker, E. A. (1968). Isolation of plant cuticles with zinc chloride-hydrochloric acid solution. *Plant Physiology, Lancaster* 43, 1878–9.

Holloway, P. J. and Challen, S. B. (1966). Thin layer chromatography in the study of natural waxes and their constituents. *Journal of Chromatography* 25, 336–46.

Hull, H. M., Went, F. W. and Bleckmann, C. A. (1979). Environmental modification of epicuticular wax structure of *Prosopis* leaves. *J. Arizona–Nevada Academy of Science* 14, 39–42.

Jeffree, C. E., Baker, E. A. and Holloway, P. J. (1975). Ultrastructure and recrystallization of plant epicuticular waxes. *New Phytologist* 75, 539–49.

Juniper, B. E. (1960). Growth, development and effect of the environment on the ultrastructure of the leaf surface. *Journal of the Linnean Society (Bot.)* 56, 413–19.

Juniper, B. E. and Bradley, D. E. (1958). The carbon replica technique in the study of the ultrastructure of leaf surfaces. *J. Ultrastructure Research* 2, 16–27.

Juniper, B. E., Gilchrist, A. J. and Robins, R. J. (1977). Some features of secretory systems in plants. *Histochemical Journal* 9, 659–80.

Kay, Q. O. N., Daoud, H. S. and Stirton, C. H. (1981). Pigment distribution, light reflection and cell structure in petals. *Botanical Journal of the Linnean Society* 83, 57–84

Kerner, A. (1878). *Flowers and their Unbidden Guests.* Kegan Paul & Co., London.

Killham, K. and Wainwright, M. (1981). Microbial release of sulphur ions from atmospheric pollution deposits. *Journal of Applied Ecology* 18, 889–96.

Kloepper, J. W., Leong, J., Teintze, M. and Schroth, M. N. (1980). Enhanced plant growth by siderophores produced by plant growth-promoting rhizobacteria. *Nature* 286, 885–6.

Knaggs, N. S. (1947). Adventures in Man's First Plastic. *The Romance of Natural Waxes*. Reinhold, New York.

Kolattukudy, P. E. (1970). Plant waxes. *Lipids* 5, 259–74.

Kolattukudy, P. E. (1976). *Chemistry and Biochemistry of Natural Waxes*. Elsevier, Amsterdam, Oxford, New York.

Lee, B. and Priestley, J. H. (1924). The Plant Cuticle I. Its structure, distribution and function. *Annals of Botany* 38, 525–45.

Lipetz, J. (1970). Wound healing in higher plants. *International Review of Cytology* 27, 1–28.

Lloyd, F. E. (1942). *The Carnivorous Plants*. Chronica Botanica Company, New York.

Martin, J. T. and Juniper, B. E. (1970). *The Cuticles of Plants*. Edward Arnold, London.

Meidner, H. and Mansfield, T. A. (1968). *Physiology of Stomata*. European Plant Biology Series, McGraw-Hill, London.

Monteith, J. L. (1973). *Principles of Environmental Physics*. Edward Arnold, London.

Mooney, H. A., Gulmon, S. L., Ehleringer, J. and Rundel, P. W. (1980). Atmospheric water uptake by an Atacama Desert shrub. *Science* 209, 693–4.

Muller, C. H. (1966). The role of chemical inhibition (allelopathy) in vegetational composition. *Bulletin of the Torrey Botanical Club* 93, 332–51.

Neilson-Jones, W. (1969). *Plant Chimeras*. Methuen & Co., London.

Nutman, P. S. (ed.) (1976). *Symbiotic Nitrogen Fixation in Plants*. International Biological Programme No. 7, Cambridge University Press.

Parsons, E., Bole, B., Hall, D. J. and Thomas, W. D. E. (1974). A comparative survey of techniques for preparing plant surfaces for scanning electron microscopy. *Journal of Microscopy* 101, 59–75.

Philips, J. (1928). Rainfall interception by plants. *Nature* 121, 354–5.

Robins, R. J. (1976). The nature of the stimuli causing digestive juice secretion in *Dionaea muscipula*, Ellis (Venus's Flytrap). *Planta* 128, 263–5.

Ruinen, J. (1975). *Nitrogen Fixation by Free-living Micro-organisms*. Ed. W. D. P. Stewart. International Biological Programme No. 6, 85–100, Cambridge University Press.

Satina, S., Blakeslee, A. F. and Avery, A. G. (1940). Demonstration of the three germ layers in the shoot apex of *Datura* by means of induced polyploidy in periclinal chimeras. *American Journal of Botany* 27, 895–905.

Schmidt, A. (1924). Histologische Studien an phanerogamen Vegetationspunkten. *Bot. Arch.* 8, 345–404.

Schönherr, J. (1976). Water permeability of isolated cuticular membranes: the effect of cuticular waxes on diffusion of water. *Planta* 131, 159–64.

Silvester, W. B. (1976). *Symbiotic Nitrogen Fixation in Plants*. Ed. P. S. Nutman. International Biological Programme No. 7, 521–38, Cambridge University Press.

Sutcliffe, J. F. (1979). *Plants and Water*. 2nd Edition. Studies in Biology no. 14. Edward Arnold, London.

Sutherst, R. W., Jones, R. J. and Schnitzerling, H. J. (1982). Tropical legumes of the genus *Stylosanthes* immobilize and kill cattle ticks. *Nature, London* **295**, 320–21.

Swart, H. J. (1972). Australian leaf-inhabiting fungi II Two new ascomycetes. *Transactions of the British Mycological Society* **58**, 417–21.

Tinsley, T. W. (1953). The effect of varying the water supply of plants on their susceptibility to infection with viruses. *Annals of Applied Biology* **40**, 750–60.

Went, F. (1974). Reflections and speculations. *Annual Review of Plant Physiology* **25**, 1–26.

Wettstein-Knowles, P. von and Netting, A. G. (1976). Composition of epicuticular waxes on barley spikes. *Carlsberg Research Communication* **41**, 225–35.

Williams, K. and Gilbert, L. (1981). Insects as selective agents on plant vegetative morphology: Egg mimicry reduces egg laying by butterflies. *Science* **212**, 467–9.

Winkler, H. (1907). Über Pfropfbastarde und pflanzliche Chimaeren. *Berichte der Deutschen botanischen Gesellschaft* **25**, 568–76.

Yeoman, M. M. (1982). Cellular recognition systems in grafting. *In:* Intercellular Interactions. (Eds Linskens, H. F. and Heslop-Harrison, J. W.). Encyclopedia of Plant Physiology, New Series. Springer-Verlag, Berlin, Heidelberg, New York. In Press.

Yeoman, M. M. and Brown, R. (1976). Implications of the formation of the graft union for organization in the intact plant. *Annals of Botany* **40**, 1265–76.

Index